NUCLEAR RENAISSANCE
Technologies and Policies for the Future of Nuclear Power

NUCLEAR RENAISSANCE
**Technologies and Policies
for the Future of Nuclear Power**

W J Nuttall
Judge Institute of Management
and Cambridge University Engineering Department

With a foreword by the Rt Hon. Brian Wilson MP

I*o*P
Institute of Physics Publishing
Bristol and Philadelphia

British Library Cataloguing-in-Publication Data

A catalogue record for this book is available from the British Library.

ISBN 0 7503 0936 9

Library of Congress Cataloging-in-Publication Data are available

For updates, discussions and comments on the issues raised by this book please visit http://bookmarkphysics.iop.org/bookpge.htm?&isbn = 0750309369

Commissioning Editor: John Navas
Editorial Assistant: Leah Fielding
Production Editor: Simon Laurenson
Production Control: Sarah Plenty
Cover Design: Victoria Le Billon
Marketing: Louise Higham, Kerry Hollins and Ben Thomas

Published by Institute of Physics Publishing, wholly owned by The Institute of Physics, London

Institute of Physics Publishing, Dirac House, Temple Back, Bristol BS1 6BE, UK

US Office: Institute of Physics Publishing, The Public Ledger Building, Suite 929, 150 South Independence Mall West, Philadelphia, PA 19106, USA

Typeset by Academic + Technical, Bristol
Printed in the UK by MPG Books Ltd, Bodmin, Cornwall

for Marion

Contents

Foreword

This book is both timely and essential. Britain urgently needs a national debate on the future direction of energy policy and, specifically, about whether or not there is to be a place for the nuclear industry. Such a debate can only take place on the basis of balanced, well-informed evidence of the kind that Bill Nuttall has provided here.

The reasons for the existence of the civil nuclear industry in this country are now part of history. The reality is that almost a quarter of the UK's electricity comes from that form of generation. In an era when unprecedented emphasis is being placed on the need for carbon reduction, the real question is not whether we can afford nuclear power—but whether, in the name of environmental responsibility, we can afford to turn our backs on it.

In order to form a view in response to that question, it is necessary to be equipped with some degree of technical understanding and recognition of the wider energy debate, as well as political opinions. Readers of *Nuclear Renaissance: Technologies and Policies for the Future of Nuclear Power* will be better qualified to look rationally at an issue which will soon have to be resolved.

Rt Hon. Brian Wilson MP

Author's notes and acknowledgments

This book has been made possible by many people and organizations. The full list of those involved from inside and outside the nuclear industry is far too long to be recorded here. I am especially grateful to my current colleagues from Cambridge University and particularly the Cambridge–MIT Electricity Project for countless insights into the workings of the modern electricity industry. Within the CMI–EP family I am grateful to my PhD student Fabien Roques for his work on security of supply and nuclear power economics. I am also grateful to my former colleagues at the Institute of Physics for giving me space when I worked for the Institute to pursue my policy interests in energy and nuclear power.

I am grateful to numerous individuals for direct assistance. These include Ralph Hall, Tom Curry and Alan Cheng from the Technology and Policy Program at the Massachusetts Institute of Technology. I would also like to thank Malcolm Grimston and the late Peter Beck for much advice and inspiration, especially in the early phases of this project. I am most grateful to Martin O'Brien and Chris Carpenter of the UKAEA, Richard Mayson and colleagues at BNFL, Rudy Konings of the EC–JRC Institute for Transuranium Elements and Jerry Hopwood and Michael Ivanco of AECL for their advice and assistance. Readers should note that, despite receiving much assistance, the content of this book is my responsibility and it should not be assumed that those who have assisted me would agree with what I have said. This book covers a wide range of topics and issues in a fast moving field. Errors and omissions are inevitable, and for these I apologize. All such failings are mine alone.

I am most grateful to the Commissioning Editor of this project, John Navas, for his support and encouragement. I would also like to thank Andrew Giaquinto and his colleagues in the IOPP Studio for their work on the cover art and the IOPP production team for being so expeditious and careful. I would like to thank Rt Hon. Brian Wilson MP for generously agreeing to provide the Foreword. I am also most grateful to IOPP's anonymous referees for their supportive words of criticism and to my parents for their attempts to teach me English grammar.

Lastly I would like to record my gratitude to my daughter Marion to whom this book is dedicated and to my wife Alisande for her tolerance. They have both put up with me being a very absent-minded academic these last several months.

W J Nuttall
Cambridge, October 2004

Chapter 1

The nuclear renaissance—an introduction

So nuclear power is coming back? Increasingly the media in Europe and North America are giving that impression. What is driving this new interest? What might nuclear power look like in the future? These are the central questions motivating this book.

Fundamental to the renewed interest in nuclear power is an increasing concern for global climate and the increasing realization that mankind needs to reduce emissions from the burning of fossil fuels, such as coal, oil and gas. Also forcing renewed interest in nuclear power is a concern that the fossil fuels, on which so much of our modern energy needs depend, come from politically unstable parts of the world. After 11 September 2001, there is also an increasing awareness that the logistics systems underpinning the movement of fossil fuels are complex and vulnerable. As this book makes clear, nuclear power has had its share of setbacks and failures. It is still perceived by many to be economically uncompetitive, but the world is changing fast. Energy policy makers are looking afresh at a range of options. Each option has its benefits and disadvantages. Nuclear power is unusual among electricity technologies in being a mature industry well placed to help with climate change and security of energy supplies. Furthermore it is possible that it is indeed an industry that has learned from its mistakes.

This book considers the possibility that nuclear power will indeed undergo a renaissance. Such a premise inevitably implies emergence from a previous period of decline and so the title of this book implies a certain perspective. The idea of fluctuating fortunes for nuclear power implies an observer with a North American or Western European perspective on such matters. By contrast, in Asia the expansion of nuclear power has progressed continuously through the past three decades and, from an Asian perspective, the notion of a nuclear renaissance is of little relevance to either technology or policy, save perhaps for scenarios concerning the future price of current nuclear power's fundamental fuel, uranium.

It is in the United States that the concept of nuclear renaissance is perceived most clearly, and it was there that the term was first coined. In 1990 *US News and World Report* ran a short story by Charles Venyvesi under the headline 'Nuclear Renaissance' [1].

However, the idea of nuclear renaissance failed to catch the public imagination in the early 1990s. Then on 13 September 1999 the idea of nuclear renaissance was repeated in an article by Mark Yost in the *Wall Street Journal* [2]:

> *Not long ago, nuclear energy looked headed for extinction. Those days are over. With production costs dropping and regulations for fossil-fuel-burning plants rising, there's a renaissance taking place in nuclear power that would have been unthinkable five years ago.*

This time the idea took hold.

The United States pioneered nuclear power in all its forms. With international assistance, it developed the atomic bomb in the 1940s. It pioneered the use of nuclear reactors in naval propulsion and it developed a large-scale commercial civilian nuclear power industry in the 1960s and 1970s. In those early years nuclear power felt good. America's pre-eminence in all things nuclear was a source of national pride and self-confidence. In nuclear technology, at least, the United States led, and the rest of the world followed. However, as we shall see, these early years of hubris and optimism were followed by a period of national self-doubt, growing fears for safety and the environment and a strong desire to question authority in all its forms. All these factors would combine to undermine civilian nuclear power.

The previously attractive attributes of the technology (engineering-led design pushing forward the limits of the possible, a culture of decision-making by experts, and the idea that man should harness nature) increasingly came to be regarded by mainstream society as being part of the problem of modern living rather than part of the solution to our collective needs.

If civil nuclear power is undergoing, or is about to undergo, a renaissance, what will history judge to have been the low point in its fortunes? Will it be the accident at Three Mile Island in Pennsylvania USA in March 1979? Or will it be the disaster at Chernobyl in the Ukraine in April 1986? In this author's opinion, while these events were indeed terrible and indicative of some of the worst problems of the industry, the true nadir of nuclear power came slightly later, in the 1990s.

In Britain and North America the 1990s were characterized by a desire to deregulate, restructure and where necessary, privatize energy utilities. This brought the discipline of the market fully to bear on energy engineering. The days of engineering-led design were gone, along with the cost plus contracts. The 1990s represented the triumph of the economists over the engineers. It would not be until the 21st century that the economic basis of the electricity industry would start to be adjusted to take account of externalities and other market failures. These externalities, such as the impact of electricity generation on global climate, could, when internalized, greatly assist the economic fortunes of nuclear power. However, in the early 1990s, when only the first steps towards liberalized electricity markets were being taken, forward-looking energy analysts would probably have predicted a bleak future for nuclear power and particularly new build.

Below we shall consider one particular story from the darkest days of nuclear power. It would be incorrect to assert that the difficulties described below were caused by the move to liberalize electricity markets, as the seeds of the story significantly pre-date such market developments. The issue is more that the story set out below has its key developments occurring in parallel with the move towards liberalized markets. Perhaps the timing and the economic context are more important than the specifics of one particular emblematic project. It is human nature, however, to seek to associate broad trends with single stories. In that spirit it is probably not unreasonable to associate the darkest days of nuclear power with one story in particular.

If any single story demonstrates the descent towards the nadir of nuclear power from the 1960s through to the 1990s it is the tale of the Shoreham nuclear power plant on Long Island, New York. The downward journey took many years and had many twists and turns, but in this author's opinion it has a strong claim to have been the nadir of nuclear power.

On 26 February 1992 the United States Nuclear Regulatory Commission authorized the decommissioning of the Shoreham boiling water reactor on the north shore of Long Island in New York State [3]. A brand new civil nuclear power plant, in fact the world's most expensive commercial nuclear power plant, was to be decommissioned without ever having sold a single unit of electricity. The decision was made when the region desperately needed electrical power, the plant was complete and fully functional and the nuclear fuel had been inserted, activating the reactor core.

The story of the Shoreham nuclear power plant is so definitive in the history of nuclear power that at least two books have been dedicated to recording the sorry tale for posterity. The first of these to seem was David P McCaffrey's 1991 history *The Politics of Nuclear Power* [4] followed in 1995 by Kenneth F McCallion's *Shoreham and the Rise and Fall of the Nuclear Power Industry* [5].

The history of nuclear power is peppered by stories of cost over-runs, construction delays, safety fears, and occasional accidents. While there have been many examples of successful nuclear power plant developments (such as Beznau I and II in Switzerland operating reliably since 1969 and 1971 respectively), there have been numerous other reactor projects that have been massively over budget and behind schedule. Of all these the saga of the plans to build a nuclear power plant at Shoreham, New York, is probably the worst of all. Importantly no side emerges from the tale with its reputation unscathed. Whether one's perspective is that nuclear power was indeed the best solution for Long Island's power needs or conversely if one believes that nuclear power should never have been considered for a densely populated island, one cannot fail to be horrified by the extent of the folly that occurred.

It is not possible here to recount the whole sorry tale of the Shoreham plant, but some key aspects of the story are worth emphasizing.

The story starts in 1965 in the heady days of nuclear optimism, when the Long Island Lighting Company (LILCO) of New York, conscious of the growing population of Long Island and the concomitant demands for electricity by domestic users and industry, decided to get into the nuclear game and founded a nuclear engineering division [6]. The following year LILCO formally proposed the construction of a nuclear power plant at an estimated cost of $65–75 million with the aim of supplying power to the island by the early 1970s.

Rather than cheap power ready for the 1970s, the final reality was that 23 years later, on 28 June 1989, the LILCO stockholders would vote to accept a deal with the state under which the completed Shoreham plant would be abandoned in return for a package of financial compensation from the state [4]. The cost to electricity consumers of the misadventure would finally total an astonishing $6 billion [7]. The cost of the Shoreham's nugatory contribution to electricity generation totalled approximately one hundred the times the original budget for a machine that the early backers promised would produce useful electricity for the people of Long Island for several decades. Instead the sorry saga of the Shoreham plant left Long Island saddled with some of the highest electricity bills in the United States.

How did the Shoreham plan go so badly wrong? Citizen activism, corporate mismanagement of the project, and a regulator that cared little for the true impacts of its decisions, had all conspired to halt the operation of a complete and fully licensed modern power plant.

The management mistakes included the selection of General Electric's novel Mark II containment structure, later judged to be inadequate and requiring $100 million to be re-engineered [3]. Another engineering and management error was the initial choice of diesel back-up generators for the plant. In 1976, three back-up diesel generators were supplied by Transamerica Delavel Inc, a company new to the nuclear business. Seven years later, in August 1983, when the plant was finally approaching completion, the crankshaft of one of the three generators broke during routine testing. Inspection of all three generators revealed that they had been manufactured with crankshafts inadequate to perform routine operations [8]. Despite the fact that the fault was repaired, LILCO eventually purchased new units from Colt Industries—a regular supplier to nuclear power plants in the United States. LILCO had been seduced into purchasing the untried Transamerica Deleval equipment by the prospect of a small cost savings when compared with the cost of tried and tested competitors. There is a certain irony in the fact that penny pinching led to problems on one of the most over budget nuclear power plants in history. In fact the penny pinching was a direct cause of much of the cost overrun and it could be argued that it was an example of the old adage: 'spoiling the ship for a ha'porth of tar'. It took LILCO seven years to realize that it had a problem with the diesel generators and this delay made the problem all the more acute. As the problem was noticed towards the planned end of the project, it is estimated that it added approximately 16 months to a project that was already in serious trouble [8].

The repeated delays raised two consequential difficulties for LILCO and its contractor Stone and Webster which might otherwise have been avoided. First, the delays allowed for the imposition of ever tighter regulatory burdens. In most cases these required retroactive compliance by the still uncompleted Shoreham plant. If the plant had been completed, there would have been no requirement to comply with the new tighter regulations. The consequence of such policies is described by Kenneth McCallion as follows:

> ... in Georgia, two plants designed to be identical were scheduled for completion in 1975 and 1978. After the first was completed, however, regulatory changes required that the second must be built with thousands of pipe supports that were twice as large as those in the first unit. 'Now these plants are one foot from each other and both are operating', commented W A Widner, a Georgia Power Co. official. 'It says something about the system.... It just doesn't make any sense' [9].

The Shoreham saga, however, takes perversity of nuclear policy beyond the idea that two plants might both be licensed to operate simultaneously to include the idea that (on the basis of different safety standards) a newer safer plant could be barred from operating by a regulator, while simultaneously an apparently less safe facility was allowed to operate.

While the modern Shoreham plant was being stifled by safety regulation, in the 1970s and 1980s the United States continued to operate increasingly elderly nuclear plant that in some cases had been constructed before there had been a proper understanding of the effects of radiation on nuclear reactor components. A rational safety-first analysis would have closed the ageing plant and replaced the lost supply capacity with new plant such as that constructed at Shoreham. The actual consequence of a policy process intended to promote safety was that the (very small) risk of nuclear accident was probably increased rather than diminished.

Political pressure grew for ever-tighter regulation of nuclear new build in the wake of the Three Mile Island accident in 1979 (see chapter 3). By this time the Shoreham nuclear power plant should have been operating for several years. The reality, however, was that the Shoreham project received the full political and regulatory impact of both Three Mile Island and the later Chernobyl disasters.

A second major consequence of the project delays at Shoreham was that it allowed opponents from within the local community ample time to mobilize against the project and to plan ever more sophisticated legal attacks against it. The history of the legal arguments surrounding Shoreham is extensive. Dominant in these disputes was a lengthy tussle between Suffolk County, New York, and LILCO concerning emergency planning. (Suffolk County consists of the eastern two-thirds of Long Island. The neighbouring county immediately to the west and, therefore closer to New York City, is Nassau County.)

Legal disputes surrounding Shoreham include [10]:

- In 1982 the local authority, Suffolk County, New York, removed itself from the contract to collaborate with LILCO on emergency planning.
- By February 1983 the county had concluded that acceptable emergency planning was not possible for Shoreham given the island's geography.
- In 1984 there was a dispute between Suffolk County and LILCO concerning the level of property taxes to be paid by the utility.
- In 1986 Suffolk County passed a local ordinance prohibiting testing of Shoreham emergency plans without the prior approval of the county legislature.
- Later the same year the Suffolk County legislature sued LILCO using anti-racketeering legislation arguing that the utility had obtained electricity price hikes on the basis of submitting knowingly false information.
- Also in that year, the neighbouring Nassau County refused permission for key facilities required for Shoreham's evacuation plans to be placed on its territory.
- In late 1986 the local radio station WALK resigned as primary broadcaster of any emergency announcements.
- In 1987 two Long Island towns refused zoning/planning permission for proposed decontamination trailers.

The legal history reveals the importance of local acceptance, and particularly local government acceptance, for developments in the area of nuclear power in the US context. Many of the difficulties surrounding the Shoreham project came from a complete breakdown of trust between the utility (LILCO) and the local authority for the region (the Suffolk County Legislature). The lessons of this aspect of the Shoreham project are clear. Community engagement (coupled perhaps with a better policy balance between the local and national interests) will be vital if any American nuclear renaissance is to be possible. Such thinking leads to the suggestion that a nuclear renaissance should start with an intention to 'replace nuclear with nuclear'. It is both proper and easier to build new modern nuclear power plants in places with existing nuclear power plants reaching the end of their operational lives. Such locations have high levels of community familiarity with the issues of nuclear power.

LILCO, however, should not simply be regarded as a victim of politics and regulation. It too made several major strategic errors. Whenever it was presented with a chance to get on with the job, it reacted with the view that it should rethink the project and improve upon it. For instance, in late 1968, when the project still had optimism and forward momentum, LILCO decided to rewrite its original application for a 540 MWe unit and re-file with plans for a larger 820 MWe single unit plant. This decision alone delayed the project so long that it became the first reactor in the United States that needed to comply with the National Environmental Policy Act of 1970 [3].

The battle between the forces for and against Shoreham was so intense that neither side was prepared to give ground in the interests of sensible policy making. In 1983 at the height of the arguments over plans to evacuate the island's population in the event of a nuclear accident, LILCO pushed ahead with plans to put nuclear fuel into the reactor. This decision may in part be regarded as a perverse consequence of the regulatory distinction between operational plant and plants under construction referred to earlier. One might argue that LILCO were rational in seeking to get the plant operational. The policy problem arising from LILCO's June 1983 request to the Nuclear Regulatory Commission for a licence to run the reactor at low power for testing purposes, was that, in doing such tests, the utility caused the reactor core to become radioactive and hence massively increased the decommissioning cost in the eventuality that then unresolved policy battles would later prompt closure and decommissioning of the site. Such a prompt closure was indeed the final outcome, but it now followed the low-power testing which had greatly increased the costs of decommissioning as a result of the extra radioactivity induced.

In October 1984 the Atomic Safety Licensing Board (a regulatory panel of the Nuclear Regulatory Commission) issued a licence for low-power testing at Shoreham [11]. The intensity of the fight surrounding Shoreham in the autumn of 1984 is apparent from the language Suffolk County legislator Wayne Prospect used at the time:

> 'the NRC could try to ram through a license for Shoreham a lot sooner than previously thought.'

Separately he went on to say

> 'it is absolutely imperative that a practical plan to close Shoreham be adopted now.'

The NRC with its regulatory creep and perverse policies had done much to hinder progress at Shoreham, and yet, in this legislator's eyes, the regulator was on the side of LILCO in trying to get the plant operational as quickly as possible.

Shoreham could not have earned its title 'the nadir of nuclear power', if it had simply been abandoned in the summer of 1983 as soon as it had been completed. Nor could Shoreham claim the title if it had started to gener-ate power in the mid 1980s with the infamy of being the most expensive nuclear power plant in America. Shoreham's credentials as the nadir of nuclear power come from the fact that it was built, fuelled, tested, closed and decommissioned without a single kWh of electricity ever being sold, and all at a cost of $6 billion once the radioactive decommissioning is factored in. The opportunity cost of such expenditures, and the benefits that might have been realized if even other opportunities in the energy sector had been so funded, hardly bears consideration. Even as one increas-ingly hears talk of renaissance and new nuclear build, it may still be the case that the Shoreham debacle was the final nail in the coffin of western nuclear power. Nuclear renaissance is still not inevitable.

Before we leave the saga of Shoreham and the scars it leaves on nuclear power, it is worthwhile adding a few more bits of information to reveal the full scope of the story.

In the 1960s as the Long Island economy grew and population increased, semi-skilled and skilled industrial workers raised families, enjoyed prosperity, found stimulating employment and enjoyed a high quality of life. Long Island communities, such as Bethpage and Patchogue had easy access to the most vibrant city on earth—'the Big Apple'—via the Long Island Rail Road and the Long Island Expressway. The United States was going to the moon and Grumman of Long Island built the Eagle lander for the Apollo programme. These were the golden years for blue collar Amer-ica. The Shoreham opposition drew its energy from another facet of Long Island life, the well-to-do retirees and vacationers of the Hamptons at the eastern end of the island. Through the 1970s and 1980s the Hamptons boomed with numerous VIPs and Hollywood stars purchasing property. A strong industrial base and its thirst for nuclear power presented few benefits for such people.

No factor played a greater part in the victory of the opponents of Shoreham than the issue of evacuation of the island in the extremely remote eventuality that there would be a serious release of radioactivity from the Shoreham plant. It might even today come as a surprise to many who opposed Shoreham that throughout this period central Long Island already had an operational nuclear reactor dating from the 1960s. It was the High Flux Beam Reactor at Brookhaven National Laboratory. This Highly Enriched Uranium fuelled research reactor with its metal casing (rather than a reinforced concrete containment), always represented a seemingly greater risk of accident or sabotage than a power reactor, and yet because of regulatory details and the nature of politics, the Brookhaven Reactor rarely came up in the arguments over Shoreham in the 1980s and early 1990s. It is important to note, however, that research reactors are significantly smaller (and lower power) than reactors built to generate electricity and, as such, they contain far less hazardous radioactive material.

Once the opponents of Shoreham had won their battle, however, they turned their sights on the Brookhaven laboratory and, following foolish prioritization of safety work by the laboratory managers, the anti-nuclear activists eventually found the BNL reactor's jugular—a highly embarrassing leak of radioactive tritium-contaminated water from the reactor's fuel cooling pond. In this way they won the final battle of their long war to rid Long Island of nuclear fission. While there is much talk these days of a nuclear renaissance in the United States, no-one believes that nuclear power will ever return to Long Island, New York—home of the highest electricity bills and some of the most expensive houses in America.

References chapter 1

[1] Venyvesi C 1990 'Nuclear renaissance', *US News & World Report*, 20 August, **109**(8) 16.

[2] Yost M 1999 'The Producers—Three Mile What?' Energy Special Report, *The Wall Street Journal*, 13 September, **234**(51) 14.

[3] Wilson K 1992 'Lights out for Shoreham', *Bulletin of the Atomic Scientists*, June (as of October 2004: http://www.bullatomsci.org/issues/1992/j92/j92.wilson.html)

[4] McCaffrey D P 1991 *The Politics of Nuclear Power* (Dordrecht: Kluwer Academic Press) p 253.

[5] McCallion K F 1995 *Shoreham and the Rise and Fall of the Nuclear Power Industry* (Westport, CT: Praeger)

[6] McCaffrey D P, op. cit., p 33.

[7] Fagin D *Lights Out at Shoreham,* Newsday.com (as of October 2004: www.newsday.com/community/guide/lihistory/ny-history-hs9shore,0,563942.story)

[8] McCaffrey D P, op. cit., p 82

[9] McCallion K F, op. cit., p 15.

[10] McCaffrey D P, op. cit., pp 234–254

[11] McCaffrey D P, ibid, p 244

PART I

THE POLICY LANDSCAPE

Chapter 2

Issues in energy policy

In the first decade of the new millennium energy policy is back on the agenda. Having dominated international relations and domestic politics in the mid 1970s, energy policy quietly receded from newspaper front pages to the business pages during the 1980s and the 1990s. During these years the role of central government policy diminished as major decisions concerning energy utilities were increasingly passed to the private sector. In Britain, Margaret Thatcher's government had spent the 1980s privatizing any and all large-scale nationalized activities that had even a chance of turning a profit. Gas and telecommunications systems had been privatized *en bloc* to form new private monopolies from former public monopolies. Market forces were intended to function through contestability. New entrants would be welcomed and encouraged to undercut the incumbent player. While such privatizations were highly profitable for the Treasury, and consumer costs did fall, in general they did not yield true competition.

Within energy policy one particular aspect constitutes the central theme of this book and that is electricity (section I.3.4 is something of an exception). Policy makers tasked with delivering the Conservative Party's 1987 election manifesto pledge to privatize the UK electricity system were determined to produce real competition by breaking the monopoly of the Central Electricity Generating Board (CEGB). The thorn in the side of this policy was to be nuclear power, which because of fears over decommissioning was found to be impossible to sell to the financiers of the City of London. First the elderly Magnox reactors and later the Advanced Gas-cooled Reactors had to be removed from the privatization plans. Once this matter had been resolved, the vast CEGB enterprise and the 12 electricity Area Boards were ready to be sold to the market as Regional Electricity Companies.

In the years following the privatization of English and Welsh electricity in 1990, the two main generators National Power (now RWE npower) and Powergen (now part of E.On) have been joined by numerous merchant generators new to the UK market and often new to the electricity industry.

Important mergers and acquisitions have taken place. The French state-owned national champion Electricité de France (EdF) has entered the English and Welsh liberalized market and assumed a major role in the British electricity industry. Its wholly owned subsidiary EDF Energy has taken over three major Regional Electricity Companies (London Electricity, SWEB and Seeboard) which together serve much of southern England.

In 2004 the English and Welsh electricity system is composed of four vertically separated functions: generation, high voltage (400 kV a.c.) transmission, lower voltage distribution and supply. Of these four functions, two (generation and supply) are fully competitive. Transmission is a regulated national private monopoly (National Grid Transco) while distribution is provided by local monopolies, the Regional Electricity Companies. This situation reflects the fact that these parts of the electricity system are indeed natural monopolies. Twenty years ago, however, most observers would have regarded the entire electricity industry as being a natural monopoly.

Supply is the management of individual customer accounts. It comprises the preparation and sending of bills, marketing, branding, selling the service to new customers, and all factors relating to the public face of the industry. A fundamental part of supply is the signing-up of new customers from other suppliers. As the margins associated with electricity supply are thin, and correspondingly the savings by switching supplier are modest, it seems that it is only less affluent consumers that are motivated to switch to achieve minimum costs. More affluent consumers switch suppliers so as to combine several of their utility bills (electricity, gas, water, telecommunications etc.). They do this seeking the relative simplicity of dealing with a single company for their utility needs. Often these more affluent consumers are seeking the reliability associated with prominent brands. Whatever the customer's motivation, be it lower costs or improved customer service, nearly 40% of UK domestic customers have switched their electricity supplier in the period 1990–2002. The corresponding figure for UK business customers is about 90% [1].

While microeconomics and liberalized markets were the key issues of the 1990s, since the turn of the 21st century a wider range of electricity policy issues have achieved prominence. Gordon MacKerron of NERA introduced the author to the idea that for the UK the fundamental issues of energy policy are summarized by a triangle of issues.

I.2.1   The energy policy triangle

I.2.1.1   Economics

Economics and the operation of efficient electricity markets remain central to British national policy for electricity. In its 2003 Energy White Paper the

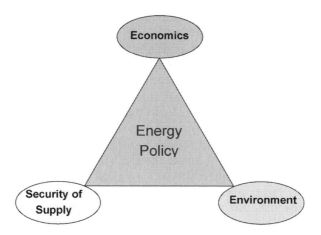

Figure I.2.1. The energy policy triangle.

Department of Trade and Industry maintained the policy that the government should not determine the mix of primary fuels for electricity generation. Renewables and their 'obligations certificates' are interestingly something of an exception to this philosophy. The British government's policy implies, for instance, that the proportion of gas-fired generation is a matter for the private sector to determine, subject to its assessment of the business and regulatory issues facing the industry. A second exception to a wholly market-led policy is nuclear power. As Gordon MacKerron of NERA has pointed out, nuclear power is regarded by policy makers as special. It needs to become ordinary [2]. Indeed, the DTI White Paper of 2003 asserts that no nuclear power plants can be constructed in the UK without a new White Paper from government. If it is the private sector that builds power plants in the UK, why is a new formal government policy document (such as is usually used to lay the foundations of new primary legislation) required before a new nuclear power plant could be constructed in Britain?

In 1990 planning permission was awarded for a new pressurized water reactor at Hinkley Point in the west of England [3]. It was to be the third reactor built on that site. The Hinkley Point C project was cancelled because of the uncertainty surrounding the creation of a new electricity market in 1990. As discussed earlier, nuclear power was proving to be an unacceptably bitter pill for the electricity market to swallow. Up until privatization, Hinkley Point C would have been a public sector project in the best traditions of the Central Electricity Generating Board. No White Paper from government would have been required for such a 'next in the series' activity. Following privatization, Hinkley Point C's future became progressively more uncertain and the whole issue became entangled with the attitude of the government to nuclear electricity. What, however, has changed fundamentally since the 1980s? At that point it was possible to build a new nuclear power plant without a White Paper. It seems that the reasons behind the government's 2003 statements concerning a new White Paper are two-fold. First the insistence on a White Paper assured all stakeholders that there would be no chance of a nuclear renaissance before the next British General Election. This yielded the government valuable political space on a contentious issue at a time when the War on Terror was taking its toll of government energy. Second, the government seems to have been convinced that new nuclear build is in fact impossible without the injection of direct public subsidy, or risk guarantees, at a level that would require a full parliamentary process and a new White Paper; but, as we shall see, this aspect of the economics of nuclear power is by no means clear.

Clearly for future nuclear power there is more in play than simply neo-classical microeconomics. In fact the once dominant position of market competition over other aspects of British electricity supply policy is now being challenged. For instance, the security of the electricity system is, at the start of the new century, a matter of growing concern.

I.2.1.2 Security of supply

The dominant issue in late 1970s energy policy was security of supply. Following the OPEC-led oil shocks of the early part of the decade, policy makers for energy and transport were acutely concerned for the stable availability of primary fuels. In those years coal-fired generation and hydro-electricity had dominant roles in electricity production. Before the invention of the combined cycle gas turbine (CCGT) and the availability of large quantities of gas from the former Soviet Union, gas was an unattractive option and the bulk of remaining generation capability came from oil-fired power stations or nuclear plant. In fact up until 1990 it was illegal to burn gas for large-scale electricity production in the European Union [4]. Following the oil crises of the 1970s, expectations (which in fact turned out to be unfulfilled) of continuously high oil prices caused the fading-out of oil-fired generation and favoured the growth of nuclear power through the late 1970s and early 1980s.

With the growth of gas-fired CCGT generation in the 1990s the range of primary fuels contributing to the UK electricity supply diversified considerably. Similar diversification has also occurred in the US although the decline of coal in favour of natural gas has been less dramatic than in the UK. As we shall see, diversity of primary fuel supplies is a key element in ensuring security of the electricity supply.

It is important to recognize that, fundamentally, electricity consumers do not want electricity—what they want are the services and comfort provided by electricity. Electricity is a means to obtain heating, lighting and communication. In the short term the use of more efficient lighting systems reduces consumers' electricity bills and improves the margin of spare capacity in the electricity generating system. However, a liberalized electricity market will slowly adjust to the changing balance of supply and demand in the system. Surplus generation capacity will gradually be eliminated from the market as generators mothball and shut down plants that are rarely required, and hence that rarely generate an income stream for the company. In this way the level of excess capacity over demand can be expected to return gradually to approximately its original level. Therefore, once the market is back in equilibrium, efficiency will have done nothing in itself to improve overall security of supply. In positing such ideas I am conscious that the very notion of an equilibrium in electricity markets might strike some as being odd. These markets have been characterized by continuous change and development and, as such, one might argue that such dynamic effects would swamp all aspects of improved energy efficiency. The incentives in the UK New Electricity Trading Arrangements (NETA) for UK generators to remove rarely used generation capacity are clear. Previously regulators mitigated this behaviour through the use of so-called capacity payments but these were removed when NETA was created, as it had been found to be a policy open to abuse.

If the electricity company has to shut down power supplies to your home, or if a tree falls on your power lines, it does not matter whether your house has old-fashioned tungsten filament light bulbs or more efficient fluorescent lighting; without power you are without light. In a liberalized free market it is likely that, in the long term, increased consumer efficiency will do little or nothing to improve the reserve capacity margin or the reliability of network infrastructures. For assured security of supply one needs, first, to minimize the risk that power companies temporarily withdraw electricity supply from their customers and, second, to improve the capacity of power companies to rectify rapidly technical faults with the electricity transmission and distribution infrastructure.

The attribute of electricity that makes it so susceptible to interruption is the requirement that supply of power from the generators must (almost) exactly and instantaneously match demand from the consumers. The larger the network or grid the more easily this is done as both generators and consumers are aggregated and fluctuations form smaller percentages of the total supply. Electricity companies are regulated with regard to the quality (voltage, frequency smoothness etc.) of the electricity they supply. There is therefore little scope to adjust the amount (voltage) of electricity supplied to consumers when supply and demand are out of balance. A serious shortage of supply with respect to demand can cause unacceptably low supply voltages (brownouts) or even force grid operators to disconnect consumers from the system (blackouts).

In England and Wales the electricity grid is buffered against mismatches in supply and demand in three key ways. First, at any instant there are spinning reserve power stations set with their turbines running at idle, but with little electricity being produced, ready to provide supply at short notice. Second, the grid also has pump storage, notably at Dinorwig and Ffestiniog in Wales. This allows surplus electricity to be used to pump water up to the top of a mountain where it is held in a reservoir (where it is also beneficially assisted by the addition of rainwater) until the point at which electricity demand exceeds supply and the water is allowed to rush down pipes within the mountain to the turbines that were earlier used to pump the water uphill. When driven by the descending water the pump storage system effectively becomes a hydroelectric power station connected to the grid. In principle, countries such as Norway with large hydroelectric capacity find it extremely straightforward to balance supply and demand on their electricity grids as hydroelectric systems are very well suited to load following operation. Interestingly in recent years drought and fears of climate change have shaken Norway out of its traditional complacency concerning security of electricity supplies.

For technical reasons, some types of generation, such as nuclear power are largely unable to operate in such a load-following mode and hence are best left to run at a continuous level irrespective of consumer demand. This type of operation is termed 'base load'.

The actual merit order of power plants from base load to peaking power is a function of both economics and technology. The English and Welsh electricity grid is protected against mismatches in supply and demand via connections to neighbouring electricity grids. This has a similar benefit to that gained by enlarging the size of an electricity grid to dampen fluctuations and to provide for increased stability but, as we shall see, it can avoid the challenging technical requirements of ensuring uniform alternating current (a.c.) synchronization across very large networks. The main connections for the English and Welsh grid are to Scotland and to France. Generally both these countries tend to have an electricity surplus and the usual mode of operation is that the English and Welsh grid buys power from its neighbours. It is interesting to note that these interconnectors are not high voltage a.c., requiring common frequencies, but rather are provided using high voltage direct current (d.c.). At such very high voltages the losses with d.c. are surprisingly low and power can be transmitted efficiently over many tens of kilometres. The technical challenges in such a system relate to very high voltage a.c.–d.c.–a.c. conversion. The ability to produce high voltage a.c. for transmission was one of the key attributes that led to the triumph of a.c. electricity networks over their d.c. competitors in the late 19th century. High voltage d.c. only became a viable possibility in the latter half of the 20th century as electrical engineering improved.

Over the years the French interconnector has become especially important for the English and Welsh electricity system. The 2 GW 270 kV d.c. interconnector with France on average operates at approximately 90% of maximum capacity with the entire flow of power being from France to

England. In the midst of the summer heatwave of August 2003, when French nuclear power plants were forced to shut down because of their impact on the temperature of river water, the interconnector with England was used for the first time to supply electricity to France. Importantly this and other interconnectors in Europe are now sufficiently congested that they add little to security of electricity supplies. Rather than interconnectors being a useful contingency for difficult days, the electricity system has come to rely upon their existence.

Pumped storage systems for electricity are dependent on suitable and increasingly hard to find geography. In addition such systems are extremely costly and burdensome to construct. These limitations led to much interest in novel forms of large-scale electricity storage including batteries and regenerative fuel cells [5]. Sadly the pioneering work of Regenysys on large-scale regenerative fuel cells for electricity storage at Little Barford near Cambridge in England has been halted on economic grounds [6]. The capacity of such storage systems to absorb or release electricity represents only a tiny fraction of generation or load on the system, and without continued research and development things will probably stay that way.

Two considerations are fundamental to reducing the risk that grid operators might find themselves having to remove supply from consumers. The first is a high degree of diversity in primary fuels used for generation, such that the interruption or high cost of one fuel type has little or no effect on the availability of the others. The second factor is an appropriate capacity margin in the system.

The reserve capacity margin or plant margin of a regional electricity system in the UK is the percentage by which installed generation capacity of power plants exceeds average cold-spell demand from all consumers. Back in the days of the Central Electricity Generating Board the target was always a 23% plant margin. This was estimated to provide for a supply greater than demand throughout all but nine winters each century. Following privatization in 1990 continuing concern for security of supply caused the creation of a special mechanism by which utilities were paid to hold spare capacity in reserve. It was an imperfect policy open to abuse and in March 2001 the New Electricity Trading Arrangements did away with the problematic capacity payments. The consequence, however, has been that plant margin estimates in the UK have decreased from an over-generous 31% in 1990 to just 16.5% in 2003, although it can be argued that, because the capacity margin has since improved, the electricity market is indeed functioning in a manner consistent with long-term security of supply.

As concern grew that the UK liberalized electricity market was failing to ensure sufficient plant margin a wave of power blackouts occurred in North America and western Europe in the late summer of 2003. Few if any of these blackouts can be straightforwardly attributed to insufficient generating capacity. Rather they flag up the dangers of poor communications between interconnected networks, legacy maintenance errors, and insufficient redundancy in network infrastructures. While diversification of primary generating fuels and generous capacity margins seem necessary for robust and reliable electricity supplies, they are not sufficient. There are dangers in regarding such system attributes as being sufficient to ensure reliable supplies.

The UK Energy White Paper of 2003 has heralded a change in terminology relating to electricity supply. The 1970s terminology of security of supply was largely replaced by the narrower concept of reliability. (Reliability excludes strategic risks such as international fuel supply disruption or major domestic industrial actions.) This evolution of terminology is a beneficial step as it can help avoid confusion with the increasing use of the term security in the context of anti-terrorism protection following the terrorist attacks on the United States in September 2001. Increasingly concerns for homeland security are impacting on the design of utility networks and power stations. These concerns are most intense in the context of nuclear installations associated with power generation and the nuclear fuel cycle. Any nuclear renaissance in Europe and North America will inevitably be shaped by concerns for homeland security.

Despite the increasing concern for homeland security and the renewed interest in reliability of electricity services, it is the environment that has emerged as the dominant issue in UK national energy policy in recent years.

I.2.1.3   Environment

While the issue of global climate change has forced environmental concerns to the forefront of British energy policy, the association of energy and environment is not new. In fact the association is as old as the electricity industry itself. Initially electricity was presented as the clean modern alternative to smoky town gas for domestic and municipal lighting. Much of the early large-scale generation of electricity was achieved with major hydroelectric projects. To a modern eye it seems amazing that some of these projects were ever permitted. For instance the construction of the Hoover Dam (originally the Boulder Dam) required the flooding of a significant fraction of one of the world's greatest natural wonders—the interlinked canyons of Arizona and Nevada, which include the world-famous Grand Canyon. Gradually concern has grown as to the impact of large-scale hydro projects on the natural environment and on nearby communities. It now seems most unlikely that any such new large-scale hydroelectricity projects will be permitted in the west, even in the unlikely event that suitable sites could be found. The most ambitious, and perhaps the world's last such project, is the vast Three Gorges Project in China.

In the 1970s and early 1980s a new environmental concern started to surface associated with electricity production. This was prompted by the death of trees in European woodlands including, importantly, the Black

Forest in southern Germany. The cause was identified as acid rain caused by sulphur dioxide and nitrogen oxides (SO_2 and NO_x) primarily emitted by coal-burning electricity power stations. The response was regulatory intervention at a European level and these emissions were successfully brought under control. European Union concern for acid rain continues to this day with ever-tightening directives on power plant emissions. In response to similar concerns, the United States pioneered the use of market mechanisms to effect environmental improvements. The so-called 'cap and trade' system allowed participants with an emissions problem to choose to purchase permits for continued emissions from those better placed to achieve inexpensive emissions reductions. This released the purchaser from the obligation to make emissions reductions directly. The consequences of such cap and trade approaches have been to achieve emissions reductions equivalent to those obtainable in more traditional command and control regulation but at a significant cost saving to the participants.

In the 1950s the new atomic power was presented as a clean pollution-free alternative to city centre coal-fired power stations that had been partly responsible for the infamous pea-souper fogs in London and other British cities. The driver of improved urban air quality for nuclear power electricity generation has continued to this day in the Far East where nuclear power construction has accelerated during the 1990s. In Europe and North America, however, the credentials of nuclear power as the clean source of electricity were tarnished in the 1980s with increasing public concerns for the industry's radioactive wastes. These aspects are discussed in detail in chapter 4.

I.2.2   Beyond the energy policy triangle

Energy policy is an extremely multi-faceted problem and devices such as the energy policy triangle described above necessarily miss important parts of the debate. One could extend the terms of the discussion in innumerable different directions, but it is probably most helpful to consider three issues in particular.

First it is important to recognize the social factors inherent in policy for energy and electricity. In the UK these factors are often considered via the concept of fuel poverty. Not without political controversy, fuel poverty is said to occur when a household spends more than 10% of its income on domestic energy needs. Fuel poverty is sometimes said to be a particularly British concept, partly arising from the large UK stock of old and poorly insulated housing. In addition, right-wing commentators have tended to argue that there is no such thing as fuel poverty. Such voices claim that there is simply poverty and it is not sensible or helpful to categorize poverty by sector. Just as there is no such thing as food poverty or clothing poverty, so there should be no special classification of fuel poverty.

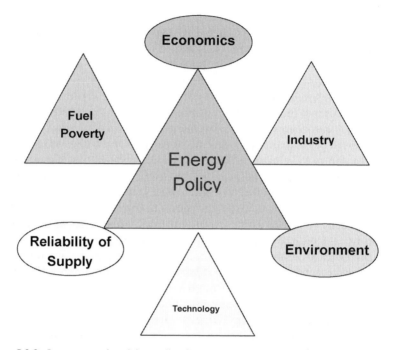

Figure I.2.2. Important electricity policy issues not captured by the bare energy policy triangle.

   Despite the controversy, the concept of fuel poverty is becoming increasingly well established in energy policy discourse. The energy uses contributing to fuel poverty comprise domestic heating, water heating, cooking, lighting and the use of electrical appliances. Many of the fuel poor are elderly and they often lack the awareness and vigour needed to replace ageing heating and lighting systems with more modern and efficient alternatives. In addition, for many a rational assessment of capital and operating costs would lead them to avoid significant investments in modern equipment, when they are of such an advanced age that they would be unlikely to live to see the full benefit of their investment. The consequence is that many old people in the UK still rely on expensive electric heating. The British pressure group The Right to Fuel Campaign is fighting to eradicate fuel poverty (see http://www.righttofuel.org.uk). This group recommends that all households should maintain a temperature of at least 20 °C in the main living room of the home and 18 °C in other parts of the home.

   The Right to Fuel Campaign's conservative estimate is that there are 4.3 million fuel poor households in the UK, while the British government figures imply a figure closer to 3 million. Whatever the true figure, it is undeniable that there is a significant set of social issues concerning electricity and energy policies in the UK.

It is interesting to note that the move towards increasingly *laissez-faire* competitive markets has been good news for the fuel poor. In the period from electricity privatization in 1990 until 2002 UK domestic electricity prices have dropped an average of 29% [1]. Furthermore the once commonplace occurrence of electricity disconnection for non-payment no longer occurs [1]. This improvement and the reduction in fear for the fuel poor are key successes for British energy policy in recent years, although the problem of fuel poverty is far from solved and prices are now rising again.

As one considers issues beyond the energy policy triangle it is important not to lose sight of an issue that was once dominant in national electricity policy-making—the relationship of electricity policy to national industrial policy. Initially the synergy of the two issues was technically motivated in that electricity systems established to provide domestic and civic lighting were well placed to use surplus capacity in the daytime to serve the needs of engineering and industry. Through the 20th century the problem became one of a system of systems. A coal-based electricity system would involve mining, transport of the coal, generation of the electricity, transmission and distribution of the electricity, and the use of the electricity by industry and commerce. As the balance of pressures within these systems shifted, one saw power stations in city centres near to railway links and later power stations constructed near mines and far from centres of electricity use and employing highly developed electricity grids and transmission systems.

Another electricity generating technology that was often bundled into issues of wider industrial policy was hydroelectricity. The World Bank and others repeatedly encouraged developing countries to embark on massive hydroelectric projects in the hope that large-scale industrialization and development would inevitably follow. The policies usually failed, however, often at great environmental and social cost to the countries concerned.

To some extent nuclear power also benefited from similar thinking: that a strong nuclear power programme in a country would give rise to a strong domestic nuclear engineering industry. Whatever the merits of such thinking, it is clearly of the past and there is no likelihood that a nuclear renaissance can sell itself on its capacity to invigorate national industrial policy, even in those countries that would still feel comfortable speaking in such terms.

One occasionally hears it said that countries such as the UK need to maintain a nuclear industry in order to retain capacity to respond to long-term global nuclear threats including terrorist fission bombs, dirty bombs and nuclear accidents in third party countries. A large nuclear industry would not be the only way to retain such capacities, and perhaps as a consequence, the argument is not widely prevalent in policy circles. Despite that, we shall return to these issues in the Afterword at the end of this book.

In most countries the only aspect of electricity with a strong residual link to industrial policy is the issue of coal. In many countries, but no longer in the UK, coal mining communities are politically active and well connected.

They are able to argue forcefully for their interests despite the, often compelling, environmental arguments against their product.

The final aspect of energy policy to be considered here is the relationship between energy policy and innovation. The relationship here is a two-way street. First, changes to energy policy and environmental regulation can spur the development of new technologies. In this context much of the work on clean coal and wind turbines seems to be being driven by energy and environment policy. Equally, however, science and engineering have the ability to generate both enabling and disruptive changes to national energy policy. For instance, the development of large-scale regenerative fuel cells has the prospect to make electricity storage a significant part of the business. If achieved this would be a truly disruptive to the orthodoxies of the electricity market. Second, developments in inexpensive plastic photovoltaic solar cell technology could pave the way for a transformation in urban architecture and design.

Another potential innovation has the potential to return the electricity back to the theme of the original great dispute between Thomas Edison and George Westinghouse in the United States: should electricity systems be based upon d.c. or a.c.? Increasingly domestic electricity use relies on 5 V or 12 V d.c. devices (flat TV screens, halogen lighting, music systems etc.). Why continue to have several tens of small inefficient transformers in your home when one larger one would suffice? Perhaps one day people will start to wire their homes with d.c. networks and electrical product designers will simplify their products and plan for such power supplies.

As Walt Patterson points out in his survey *Transforming Electricity*, these are turbulent times for the electricity business [7]. Patterson asserts that the future for electricity is not one based upon the large central generating plants and national electricity grids of the past. It is rather to be based on local distributed generating systems sited much closer to the points of electricity use and under much closer control by the user. Perhaps nuclear is inevitably a technology best suited to the outdated central plant paradigm of the past; or perhaps it will provide an essential source of robust and reliable carbon-free electricity to underpin an electricity system increasingly forced to accommodate intermittent renewables. The following chapters examine the benefits and disadvantages of future nuclear technologies in the context of a wide range of energy policy pressures.

The only certainty would seem to be that the future of electricity will not be simple and it will not be like the past.

Chapter 3

Nuclear electricity generation

In chapter 2 we surveyed the main issues in modern western energy policy with a particular emphasis on electricity generation and supply. In 2002 electricity from nuclear fission provided approximately 16% of the world's electricity needs [8]. There are wide international variations in the primary fuels used for electricity generation. Figure I.3.1 illustrates both the distribution and absolute magnitudes of electricity supplies from different fuel types in different countries. Australia and to a lesser extent the United States are strongly dependent upon coal for electricity production with all its concomitant environmental difficulties associated with emissions. The United Kingdom and Russia are noteworthy for their reliance on natural gas for electricity production. One key difference, however, between the two countries is that the UK will soon become reliant upon imported gas (including from Russian sources) while Russia will have sufficient domestic gas supplies to cover its needs for the foreseeable future. The extent to which UK reliability of electricity supply is jeopardized by the impending reliance upon gas imports is a matter of some controversy, although the British government remains relatively sanguine. As its 2003 Energy White Paper makes clear, markets can, and will, be trusted [9]:

> We do not propose to set targets for the share of total energy or electricity supply to be met from different fuels. We do not believe that government is equipped to decide the composition of the fuel mix. We prefer to create a market framework, reinforced by long-term policy measures, which will give investors, business and consumers the right incentives to find the balance that will most effectively meet our overall goals.

While nuclear power contributes 16% to global electricity needs its contribution varies widely from country to country. As figure I.3.1 illustrates, nuclear power made no contribution to Australian electricity supplies. Such a situation is more the norm than the exception, with approximately 80% of

Fuel for electricity generation (percent)

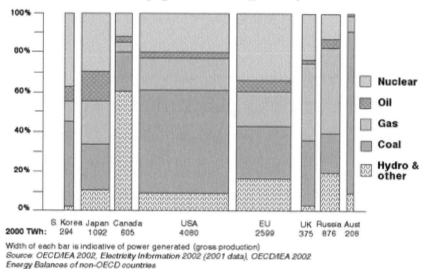

Figure I.3.1. Contribution by fuel type to national electricity generation (source: Uranium Information Centre Ltd [8]).

the world's countries making no use of nuclear power for electricity generation. By contrast France generated 77.9% of its electricity from nuclear fission in 2002 [10]. The equivalent figures for the United Kingdom and the United States of America are 24.0% and 20.3% respectively, although at the time of writing these last two figures are officially provisional and awaiting final confirmation. The figure for the United Kingdom refers to domestic generation and therefore excludes the small (~2.3%) of the total that is nuclear electricity imported from France via the English Channel interconnector (the figure of 2.3% is calculated using data from the *UK Digest of Energy Statistics 2001* [11]).

International trends in nuclear power production are also illuminating. Figure I.3.2 shows that while total electricity generation in the OECD countries continues to rise, nuclear electricity production seems to have reached a peak at approximately 2200 TWh per annum. This conclusion is, however, affected by one-off events, particularly maintenance problems with the Tokyo Electric Power Company reactors in Japan and the Advanced Gas-cooled Reactors in the UK. Figure I.3.2 shows an approximate doubling of nuclear electricity generation in the ten years 1984 to 1994, but the years 1994 to 2002 show a more modest growth. Even neglecting the particular difficulties in 2002, it is clear that nuclear power globally is not expanding at the rate it once did. Overall growth in electricity demand is also slowing, so

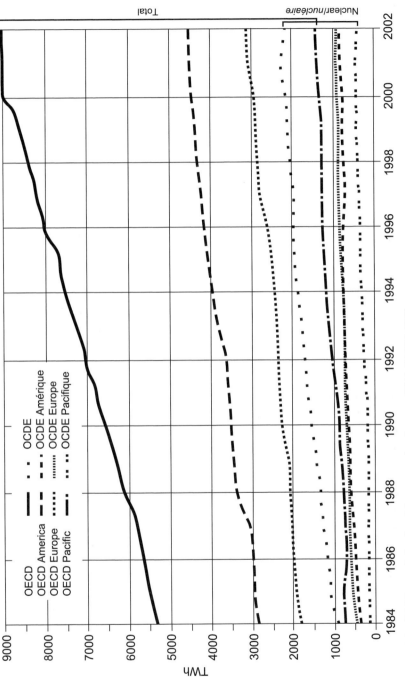

Figure I.3.2. OECD trends in electricity generation (source: NEA OECD [10]).

that proportionately the contribution of nuclear power to global electricity remains both solid and significant.

The rest of this book will be devoted to the future for nuclear power, and the technologies that would comprise any expansion of nuclear power. This chapter considers the key policy issues facing nuclear power now and in the future. However, before we discuss these issues we should consider what the technology involves and introduce some of the terminology appropriate to the field.

I.3.1   Nuclear power—how does it work?

In the late 1940s and the 1950s it was commonplace to refer to the new energy source from nuclear fission as 'atomic power'. The power of the atomic bomb was to be harnessed for peaceful purposes. While the terminology atomic is technically vague (after all, chemical energy is atomic in origin) it still provides a good first link to the workings of nuclear energy.

Atoms constitute all the material with which we interact in our daily lives. Collections of atoms bonded together are known as molecules. Water, for instance, has a molecular structure combining two atoms of hydrogen and one of oxygen. Many solids and liquids are not molecular, however. For instance metals, such as iron, consist of crystals of vast numbers of atoms held in regular arrays.

Each chemical element corresponds to a type of atom with a fixed and precise number of tiny heavy positive particles known as protons in its core or **nucleus**[1] (see figure I.3.3). In the first half of the 20th century elements were known with up to 92 protons in the nucleus. This most complex of these 'natural elements' was uranium, and it is uranium that made possible the production of nuclear energy. A key discovery of pre-war nuclear physics was that not all the mass of the atomic nucleus was provided by protons. An electrically neutral particle of similar mass to the proton and known as the **neutron** also forms part of the nucleus. Neutrons allow for a new force known as the 'strong nuclear force' binding the nucleus together. Neutrons, and the associated strong nuclear force, allow nuclei with more than one proton (i.e. all elements higher than hydrogen) to be held together despite the electrical repulsion between the positive protons. Furthermore, it was the neutron that explained why the nucleus of all atoms heavier than hydrogen had been observed to have a mass greater than the number of protons contained in the nucleus. In fact, most nuclei have a roughly equal number of protons and neutrons. In heavier atoms, however, the number of neutrons can significantly exceed the number of protons. It is the neutrons, via the

---

[1] In this section, technical terms central to an understanding of nuclear fission are shown in bold font.

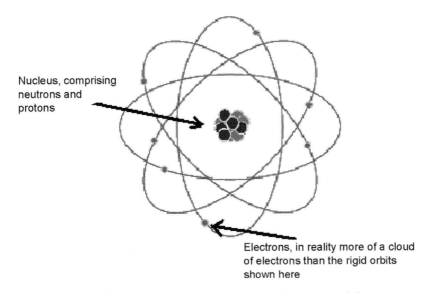

Nucleus, comprising
neutrons and
protons

Electrons, in reality more of a cloud
of electrons than the rigid orbits
shown here

Figure I.3.3. A simplified model of the atom inspired by the 'Bohr model'.

strong nuclear force, that provide the glue to hold these large, and in some
cases rather unstable, nuclei together. Hydrogen does not require neutrons
as it has only one proton and therefore no tendency for its nucleus to fly
apart—no glue is required. Despite this, there are heavier versions of
atomic hydrogen that do contain neutrons. For instance, the form of hydrogen with one proton and one neutron is known as **deuterium** and the form
with one proton and two neutrons is known as **tritium**. It is unique that
the forms of hydrogen with different numbers of neutrons (and hence different masses) have different names.

The vast majority of elements come in forms with several masses (i.e.
with several different numbers of neutrons). Each type of the same mass is
known as an **isotope**. For instance, tritium and deuterium are isotopes of
hydrogen. Isotopes have the same numbers of protons as one another,
they are different forms of the same element, and in almost all respects
they have the same chemical and physical properties as one another.
Uranium, for instance, has six reasonably stable isotopes of which three
are naturally occurring. Every uranium isotope, by definition, has 92
protons. The six possible isotopes of uranium have between 140 and 144
neutrons or 146 neutrons and the three naturally occurring isotopes have
142, 143 or 146 neutrons. Isotopes are not conventionally labelled according
to the number of neutrons, but rather according to the total number of
protons and neutrons in the nucleus. For instance, the heaviest uranium
isotope has 238 **nucleons** (92 protons and 146 neutrons) and in this book
we adopt the notation uranium-238 to denote such an isotope.

Atoms overall are electrically neutral. This results from there being an equal number of electrons in the atom to the number of positive protons in the nucleus. The electrical charge of the proton and the electron are equal in magnitude, although opposite in sign.

In 1913 Niels Bohr proposed a simple model for the hydrogen atom in which the electron orbits the atomic nucleus in a fixed circular orbit. This description of electrons orbiting the nucleus remains a useful approximation and it forms the basis of the atomic structure shown in figure I.3.3. In fact it has been determined that the electrons form shaped (spherical, figure '8' shaped and exotic) probability distributions, more reminiscent of clouds than precise planetary orbits. Of the several shapes of electron distribution the simplest is spherical. Hydrogen isotopes have spherical electron distributions and so in some ways Bohr's approximation of circular electron orbits was not so very wrong.

Fundamental to all commercial nuclear power is the process of **nuclear fission**; that is, the splitting of the atomic nucleus when it is hit by an incoming neutron. Fission occurs for only some of the heaviest elements such as uranium. It relies on the fact that the binding energy of the strong nuclear force in the largest atoms is larger than it would be in two separate atoms of roughly half the size. That is, by splitting the large, heavy uranium atom into two lighter atoms, such as barium and krypton, the difference in strong nuclear force binding energy is released. This is the energy associated with nuclear fission and all current nuclear power production.

It will be useful, however, to consider this process in more detail.

The actual measured mass of a uranium nucleus is not exactly the same as the sum of the masses of the separate protons and the mass of the appropriate number of neutrons. The measured mass is greater as the measurement of the mass actually samples the strong force binding energy. When fission occurs the incoming neutron is briefly absorbed, forming an unstable excited state. The nucleus then splits and between two and three neutrons are emitted at high speed. The combined mass of the new nuclei and fission neutrons is slightly lower than the combined mass of the incoming neutron and the original large nucleus. This measurable difference corresponds to the difference in strong nuclear force binding energy. The binding energy and the measurable mass difference are related by the famous equation first proposed by Albert Einstein in 1905: $E = mc^2$, relating energy, E, and lost mass, m.

In Einstein's equation c denotes the speed of light in a vacuum, a very large quantity, whatever one's preferred system of units (e.g. 299 793 km/s). Einstein's famous equation explains that very small changes in mass can result in very large amounts of released energy. In the fission process the energy emitted is released as kinetic energy to three types of particle, first and most intractably to extremely light and weakly interacting particles known as **neutrinos**. Neutrinos are so weakly interacting that it is extremely difficult even to detect them, let alone make use of them. Their existence,

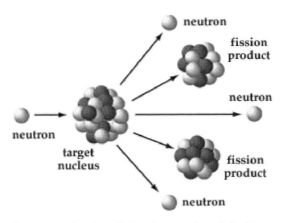

Figure I.3.4. Fundamentals of nuclear fission (source: AtomicArchive.com).

however, is beyond doubt and recently their key properties have become better understood (for a technical review of the latest in advanced neutrino physics the reader can consult the work of S M Bilenky [12], while for a more accessible overview of recent development in neutrino physics the reader is recommended to consult a recent article in *Scientific American* by Kearns and co-workers [13]). The only importance of neutrinos to this explanation of nuclear fission is that that they take away some of the energy released in the fission process. As it is so difficult to interact with neutrinos, the energy they carry is effectively lost. It does not contribute to electrical power production in any way. The two other types of particle taking energy away from the fission process are fission fragments (equivalent to the nuclei of smaller atoms) and neutrons. In each case the energy is in the form of kinetic energy of the particles ejected in the fission process. The fission neutrons carry away much of the energy, as they are travelling at about 20 million metres per second (approximately 70 million km/h). The fundamentals of the fission process are illustrated in figure I.3.4.

In effect essentially all the very heaviest elements are fissionable using these very fast high-energy neutrons known as fast neutrons. In the case of several special heavy element isotopes something special occurs if the neutrons are slowed down or **moderated**. The process by which the neutrons are slowed down should be familiar to players of table ball games such as billiards, pool or snooker. The neutron loses kinetic energy via a process of collisions. Most energy is lost if the neutron collides with a particle of similar mass, typically a light atom. The collisions are essentially elastic in that no kinetic energy is absorbed by the atoms themselves. All the neutron kinetic energy remains as kinetic energy. It is simply divided between the particles involved in the collision. A neutron colliding with a very heavy atom would transfer very little energy to the heavy atom and would simply

Radioactivity

Radioactivity is a commonplace natural phenomenon, the discovery of which predates the discovery of nuclear fission by more than 40 years. Radioactivity is harmful to living tissue and yet humans and other animals have not evolved sensory organs with which to detect the danger. It is the invisible nature of radiation that contributes much to public anxiety concerning nuclear power. Scientific instruments have, however, been developed that are sensitive to all types of radiation and these are widely used to ensure that workers and the public are not put at risk by nuclear processes.

The property of radiation that is potentially harmful is its ability to ionize atoms within our bodies; that is, to knock electrons out of atoms in our cells altering their chemical properties. In very large doses such radiation can cause tissue death. At lower levels the concern is that if the ionization occurs during the process of cell division then a mutation to the genetic programming of the cell might occur. In most cases this would also lead to cell death, but there is a risk that harmful cell mutation can survive within the body and possibly cause various types of cancer. Cancer is a particularly horrible disease and this is another reason for public concern about nuclear power.

Radioactivity is a spontaneous, natural and random process that can occur in three basic forms known as alpha, beta and gamma decay. Alpha radiation is the emission of a clump of protons and neutrons from the nucleus equivalent to the nucleus of a helium atom (two protons and two neutrons). As a consequence of its relatively large mass, an alpha particle can carry a significant amount of energy, sufficient to do serious damage to any living tissue it encounters. However, also as a consequence of its bulk, such radiation is easily blocked, such as by just a few centimetres of air or a piece of paper. The skin, which is relatively insensitive to radiation, also acts as a barrier to alpha radiation, so external sources of alpha radiation are not especially dangerous. The real risk from alpha radiation would occur if alpha emitting material were to be ingested into the lungs or the digestive system.

Beta radiation consists of very fast moving electrons. Its potential for biological harm is lower than with alpha radiation, but its penetrating power is somewhat greater. Beta radiation can be screened by a sheet of Perspex or a metal foil. Without such screening, a substantial external source of beta radiation would be serious cause for concern.

Gamma radiation is a very high-energy form of light. Unlike the other two forms of radiation it is not electrically charged and it has no mass. It is basically a very high-energy x-ray. While fundamentally not as biologically damaging as either alpha or beta radiation, gamma radiation is a particular concern as it is extremely penetrating. Several centimetres of lead or several metres of concrete are required to fully shield gamma radiation.

The idea that we are vulnerable to differing forms of radioactivity (requiring, in some cases, substantial shielding), that we cannot detect it without artificial devices and that it has the potential to damage our genes is somewhat un-nerving. The reality, however, is that mankind has always lived with radiation and radioactivity. Radiation is all around us. It comes from outer space. Radioactivity is in the rocks beneath our homes, and it is in our drinking water. It always has been and always will be. X-rays are similar to gamma rays and we routinely volunteer to expose ourselves to intense bursts of this kind of radiation for the benefits it gives us in medicine and dentistry. These doses are enormously larger than the radiation exposure one should ever get from living near a nuclear power plant.

The key to radioactivity is probability. Even a small sample of radioactive material contains a vast number of atoms. More than we can even imagine. For a radioactive material each atom has a probability every second of decaying releasing a radioactive particle. For natural materials these probabilities are tiny, but because of the large numbers of atoms involved the effect is noticeable. For radioactivity all atoms of a given isotope are equivalent and each has the same small chance of decaying. This means that the amount of radioactivity in a sample is proportional to the number of atoms of that given isotope in the sample. Once an alpha or beta radioactive decay occurs the atom will change from one isotope to another. Let us assume that the isotope produced is not itself radioactive. This means that each radioactive decay reduces the number of atoms of the radioactive isotope remaining to decay. In mathematical terms such a process is described as follows.

If N is the number of atoms of the radioactive isotope in the sample and the Greek letter λ defines the random probability of a radioactive decay then

$$dN/dt = -\lambda N.$$

This implies that if the sample has initially N_0 atoms of the radioactive isotope, then

$$N = N_0 \, e^{-\lambda t}.$$

This in turn makes clear that the time for exactly one half of the initial number of radioactive atoms to decay is a constant, whatever the starting number. This time is known as the **half-life** $T_{1/2}$:

$$T_{1/2} = \ln 2/\lambda = 0.693/\lambda.$$

Just as the probability of radioactive decay is fixed for a given radio-active isotope, then so is the half-life.

There are numerous units in circulation associated with radio-activity and its effects. In this book we follow the units specified in the primary sources quoted. In order to cross compare, it is worth noting that one becquerel (1 Bq) is equivalent to one disintegration per second, and that 1 curie (1 Ci) equals 3.7×10^{10} Bq.

bounce off with very little deceleration. In nuclear reactors fast neutron moderation is usually achieved using water (in which the hydrogen is the most important component) or carbon in the form of graphite. Graphite and water are stable enough and sufficiently easily handled to be good engi-neering materials for nuclear reactor design. Several reactor designs use so-called **light water** as the moderator (that is water with conventional isotopic hydrogen) whereas others, such as the Canadian CANDU design (see chap-ter 5), use so-called **heavy water**. Heavy water is water manufactured using deuterium instead of normal isotopic hydrogen. These issues are discussed further in chapter 5, where various water-cooled reactor designs are discussed in depth.

The process of moderation involves decelerating the neutrons to approximately one-thousandth of their initial speed. Once decelerated, the kinetic energy of the neutron compares with the kinetic energy of particles vibrating with simply their thermal energy (for a thermal neutron this is an energy of about 25 meV). As the neutron energy is comparable with thermal energies, the neutron, once moderated, is said to be **thermal**. As alluded to earlier, thermal neutrons are special because they are able with high efficiency to fission a class of nuclei know as **fissile nuclei**.

The fissile nuclei are all very heavy elements with an odd number of protons and neutrons in the nucleus. Important examples include uranium-235 and plutonium-239. The key difference between thermal neutron fission of fissile isotopes and fast neutron fission of the same isotope lies in the probability that, when the neutron strikes the target nucleus, fission will occur. A neutron of thermal energy is approximately 1000 times more likely to cause the fission of uranium-235 than its fast neutron analogue. While thermal neutron fission is technically far more attractive from both physics (easier for the reasons described above) and engineering perspectives (lower temperature reactor core) the difficulty of the method lies in the fuel preparation and manufacture. Historically civil nuclear power has been developed based upon the fission of uranium-235. While this isotope is a natural component of uranium as mined, it represents only approximately 0.7% by weight of all the uranium isotopes. By far the largest part of natural uranium is the non-fissile isotope uranium-238, which makes up 99.3% of natural uranium. While it is only the uranium-235 that is fissioned in the nuclear energy producing process, the fuel for **thermal reactors** does not need to be 100% fissile uranium-235. In fact several reactors types have been designed that can operate with natural uranium fuels. These designs include the early British gas-cooled design known as 'Magnox' and the original Canadian CANDU reactor systems discussed in chapter 5.

At a first impression, atomic hydrogen would seem to be the optimal moderator as it has almost exactly the same mass as the neutron. Unfortunately, however, atomic hydrogen is not stable and gaseous molecular hydrogen would be an ill-advised alternative because of its flammability. Hence, conventional light water with two atoms of hydrogen per molecule is frequently chosen. While light water is an extremely efficient moderator it suffers from the fact that it is also an extremely effective neutron absorber; this means that light water moderated reactors, including the pressurized and boiling water reactors discussed in chapter 5, all require fuel with more fissile material than is found in natural uranium. The traditional way to increase the fissile component of uranium thermal reactor fuel is known as **enrichment**. The technology of enrichment has its roots in the history of uranium-based nuclear weapons for which so-called 'highly enriched uranium' (HEU) is required. Highly enriched uranium is regarded as being uranium enriched to the point where at least 20% of the uranium by mass is uranium-235. HEU has no uses in the civil nuclear industry, except in some research reactors where the higher neutron fluxes emitted by very compact reactor cores (made possible by the use of HEU) have been found to be useful. There are international calls for even this limited civilian use of HEU to be entirely phased out. The most highly enriched grades of HEU (those with more than 90% uranium-235) are known as weapons grade uranium. Such materials are widely considered to be a nuclear weapon proliferation hazard, while **low enriched uranium** (typically ~3% uranium-235) required by **light water reactors** is not so problematic given the extreme technical complexity in enriching from LEU to HEU. While it can be argued that HEU represents a far greater nuclear weapons proliferation hazard than plutonium, it is important to realize the lack of linkages between HEU and the civilian nuclear fuel cycle.

Over the past 60 years there have been developments and improvements in enrichment technology. The original United States efforts dedicated to the Manhattan Project (for the development of fission weapons in World War II) attempted a wide range of uranium enrichment techniques before adopting thermal liquid diffusion and electromagnetic separation (at the Y-12 Plant at Oak Ridge National Laboratory). In the years after the war electromagnetic techniques were abandoned and gaseous diffusion technology was developed based around uranium hexafluoride high-temperature gas. This technique is still used in the United States. Other techniques include gas centrifuge methods and, on a smaller scale, laser enrichment technology. This latter technique uses very high-powered carbon dioxide lasers. These ionize uranium atoms as part of a mass-spectrometer style system. Uranium enrichment is one example where previously strong technological barriers to weapons material proliferation could erode in the future as a consequence of technical innovation.

While uranium enrichment will probably continue to improve in efficiency in future years, it is already a mature industrial-scale technology, the main player in Europe being an industrial consortium known as Urenco with facilities in several European countries including the Netherlands and the UK. As the UK Magnox reactors are soon all to be decommissioned and the Canadian successor to the original CANDU concept, known as the Advanced CANDU Reactor (ACR), will use low enriched uranium fuel, it seems that the days of the natural uranium reactor will soon be behind us. It would seem that natural uranium reactors will *not* be a technology of the nuclear renaissance.

While we have considered several concepts fundamental to nuclear power, we have not yet considered the ways in which the various elements are engineered together to produce a working nuclear power plant. In chapters 5 and 6 we shall review in detail various fission reactor designs. At this point, however, it is useful to introduce reactor concepts and we shall do this with reference to one particular design concept—the gas-cooled reactor. We choose this as our reference example because gas-cooled reactors have a greater separation of core functions (cooling, moderation etc.) than most water-cooled systems.

In figure I.3.5 we show a schematic of the core of a gas-cooled reactor. This shows the uranium fuel in long **fuel assemblies**. Each fuel assembly can consist of one or more **fuel rods**, depending on the detailed design of the reactor. Between the fuel assemblies is the moderator required to slow the fast neutrons emitted by the fission process down to thermal energies. Gas-cooled reactors commonly use carbon dioxide or helium as the coolant and graphite as the moderator. Graphite is well suited as a moderator as it is a stable solid engineering material comprising atoms with low mass nuclei. Graphite is composed of weakly bound sheets of carbon atoms. Most carbon atoms have a total of only twelve neutrons and protons in the nucleus. The sheet-like structure of the material allows it to absorb the kinetic energy of the fast neutrons from the fission process. Graphite moderators can, if operated at very low temperatures, hold this energy for quite some time in a form known as **Wigner energy** (see text box for a detailed description). If the operators were not careful, a graphite moderator in this state could release energy unexpectedly in the future. Commercial reactors are not operated at these low temperatures. Wigner energy effects are now, for the most part, understood, but this issue lies at the heart of the infamous

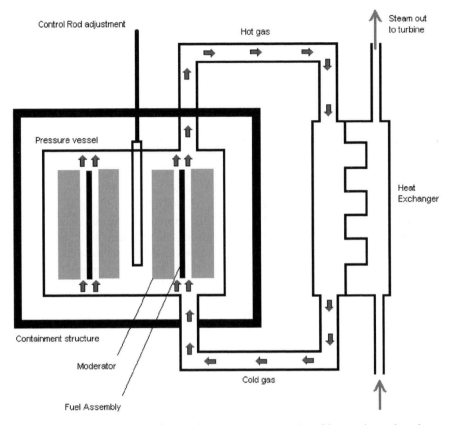

Figure I.3.5. Fundamentals of a nuclear power reactor (graphite moderated and gas cooled).

1957 fire in the plutonium producing British military reactors known as the Windscale Piles.

It is important that any power reactor remains in a state of controlled **criticality**, that is, that the numbers of neutrons emitted in the fission reaction are exactly sufficient to sustain the reaction, but not so numerous as to cause it to grow in fission and to run out of control. Management of criticality is one of the key issues in new reactor design. More than 60% of neutrons emitted by the fission process fail to cause a subsequent fission. About 23% are simply 'lost' as they are absorbed by the reactor vessel or the small atoms arising from previous fission. These smaller atoms are known as **fission products**. Roughly 38% of the fission neutrons are absorbed by the nuclear fuel without causing fission. Many are absorbed by the uranium-238 in the fuel eventually generating plutonium-239, itself a fissile isotope. Some of the neutrons are absorbed by fissile isotopes, such as uranium-235 and plutonium-239, but without causing fission. These

---

Wigner energy

The first indications that something unusual happens when low-temperature graphite is irradiated with neutrons were observations that it increases in volume or 'swells'. At higher temperatures this effect does not occur and these effects are not problematic in normal graphite-moderated power reactors such as the British Magnox and AGR designs, as these are designed to operate at relatively high temperatures in order that steam can be raised to drive the electricity generating turbines. The Windscale Piles discussed in the text, however, were not designed to raise steam—they were designed to produce plutonium by neutron irradiation of uranium-238. A low-temperature graphite-moderated reactor is very well suited to this purpose. The swelling of the graphite moderator in such circumstances was poorly understood by the British who had embarked on a crash programme of plutonium production following their surprise at being expelled from their US collaborations as a result of the US 1946 McMahon Act.

Eugene Wigner in the United States had noticed that the swollen graphite was able to store significant amounts of energy in what physicists call a metastable state. The graphite could hold the trapped energy for extremely long periods, but if it were given just a small amount of energy through external heating, the stored energy could all be released. This Wigner energy would heat nearby swollen graphite, causing yet more Wigner energy to be released and so on.

From the earliest days of interest in Wigner energy in the 1950s it has been known that the neutrons thermalizing in the moderator by collisions cause displacements of the carbon atoms of the graphite. At low graphite temperatures these displacements can stick. The carbon atoms rebind in stressed ways waiting for a bit of external heat to set them free. The precise nature of the atomic displacements remains a source of scientific interest. In 2003, for instance, a paper on just this topic was published in the elite physics research journal *Physical Review Letters* by C P Ewels and colleagues from Sussex and Newcastle Universities in the UK [46].

Despite continuing research interest in the phenomenon of Wigner energy it seems certain that it will always, above all, be associated with the Windscale fire of 1957.

---

processes of neutron absorption create the minor **actinides** that we consider in the context of nuclear waste management in chapters 4 and 7.

Another 38% of the fission neutrons do go on to produce new fissions in the fuel. It is this proportion that must be controlled precisely. As each uranium-235 fission produces on average 2.59 neutrons per fission it is important for a reactor controlled to be in criticality that 1/2.59 (i.e. 38.6%) go on to produce further fissions. Controlling the process so precisely is achieved by means of neutron-absorbing **control rods**. These can be inserted into the reactor core to soak up surplus neutrons if it is noticed that the reactor is starting to run above perfect criticality. The deeper the rods are inserted into the core, the lower the reactor criticality. If the control rods are fully inserted the nuclear reaction stops completely and the reactor is said to be **shut down**.

All modern reactor designs also include separate emergency shutdown capability in the event that primary control via the control rods is lost.

Once a reactor is shut down it does not mean, however, that it is completely benign. Of greatest concern is the **decay heat**. This source of heat persists at problematic levels for several hours after the fission reactions have stopped. It is the result of radioactive decay processes primarily in the shorter-lived fission products. Decay heat is sufficiently intense that it requires that the reactor core be cooled even after the reactor has been shut down. Decay heat is a significant concern as in the absence of any cooling whatsoever it would be sufficient to break down the integrity of both the fuel assemblies and the reactor containment. In the worst case, and in the absence of measures designed to militate against such risks, this breakdown could be sufficient to release harmful radionuclides into the environment. Some future reactor designs are being developed that do not require active cooling systems for decay heat removal (for instance, the Pebble Bed Modular Reactor discussed in chapter 6 has the potential to handle any post-shut down decay heat entirely passively). Reactors up to the present have, however, required active cooling for decay heat removal. This reality lies at the heart of several serious reactor incidents such as the 1979 Three Mile Island reactor accident in Harrisburg, Pennsylvania (discussed later in this chapter). **Loss of coolant accidents** or 'LOCAs' as they are known form a sizeable part of the discussion in Hewitt and Collier's fascinating book *Introduction to Nuclear Power*. Readers seeking an introduction to the technical realities of reactor safety are recommended to consult that text [14]. Decay heat is an example of a reactor process occurring after the main fission **chain reaction** has stopped. Processes occurring as part of the fission chain reaction described above are termed **prompt processes** and the neutrons emitted in the fission process are termed **prompt neutrons**. In addition to the prompt neutrons there are **delayed neutrons**. Delayed neutrons are fundamental to the balance of criticality within a nuclear power plant. The fact that they respond more slowly than the prompt neutrons allows a nuclear power plant to be safely controlled by gradual adjustment of the control rods.

Figure I.3.5 also illustrates the reactor shielding. In most current reactors the first shielding encountered by neutrons heading out of the **core** will be the wall of the reactor primary pressure vessel. As discussed previously, some of the neutrons will be absorbed by the walls of this **pressure vessel**. Importantly, however, some neutrons will also be reflected back into the nuclear reaction. In fact some modern reactor designs such as the Pebble Bed Modular Reactor discussed in chapter 6 rely for reactor control not on control rods inserted into the reactor core but on control rods inserted into the reflecting walls of the pressure vessel.

The coolant provides two vital functions for the reactor. The first has already been suggested: the safe operation of the reactor itself. The second is the extraction of the all-important heat energy from the reactor core in order to generate electricity. Nuclear power requires the safe, efficient production of usable heat, and the coolant system in all fission reactors is fundamental in that regard.

Figure I.3.6 shows a wider schematic of a gas-cooled reactor including the heat extraction and the production of electricity in the power station turbines. All current fission reactors raise steam in a coolant circuit, and in essence it is this steam that drives the electricity generating turbines. The turbine hall of a nuclear power plant is little different from the turbine hall of a coal-fired power station. A key difference, however, is the importance of the coolant circuit for normal reactor operations. The consequences of a major rupture of the steam pipes in a coal fired-plant, while serious, would not have the potentially dangerous consequences that a similar event might have in a nuclear plant if adequate back-up systems had not been included in the design. In addition to illustrating the cooling system of a gas-cooled reactor, figure I.3.6 also better indicates the overall structure of this particular type of nuclear power station.

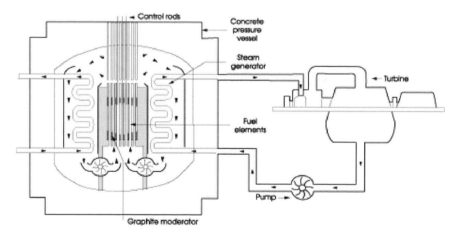

Figure I.3.6. Schematic representation of the UK Advanced Gas-cooled Reactor (source: World Nuclear Association).

While engineering and design are inevitably fundamental to the operation of nuclear power plants, the days of engineering-led-design, unconcerned for the wider contexts of cost and circumstances, are long gone. Equally fundamental to the future of nuclear power must be the economics of the technology.

I.3.2   Nuclear power economics

Following widespread liberalization of electricity markets in the 1990s, one hears repeatedly that nuclear power is uneconomic or uncompetitive. For instance the Nuclear Energy Agency has reported [15]:

> *The competitiveness of nuclear power plants has decreased substantially in recent years, particularly when compared with gas-fired plants. . . . Nuclear power is seldom the cheapest option for plants to be commissioned by 2005–2010.*

A view seems to be prevalent among opinion formers and policy makers that nuclear power is intrinsically and inevitably uneconomic. In fact, the economic fundamentals of current nuclear power plant designs are not as unsound as is often suggested. This is despite the fact that the current liberalized market framework tends to be poorly matched to the attributes of nuclear power. In this section we shall consider the position of nuclear power within current and future electricity markets.

In those countries that have moved towards market liberalization it would seem that there can be no going back. The 2003 UK Energy White Paper reports [16]:

> *Liberalized energy markets are a cornerstone of our energy policy. Competitive markets incentivize suppliers to achieve reliability.*

Before we look at the present and consider the future, it is important to recognize that nuclear power, in every country where it was developed, benefited from huge amounts of public subsidy. These subsidies acted directly for plant development and in some cases indirectly via defence expenditures for naval propulsion and nuclear weapons. Detractors make a persuasive argument that such defence-sunk costs have given nuclear an unfair advantage over alternative generation technologies. Furthermore it can be argued that this advantage persists to this day. It should be remembered, however, that gas turbine technology underpinning combined cycle gas turbine and natural gas-fired power stations is also a beneficiary of public subsidies to aeroplane engine manufacture, and hence, these gas turbine innovations also represent a military technology development.

For the most part, nuclear fission is a mature technology. While numerous innovations and improvements would surely lie ahead in the event of a nuclear renaissance, the fundamentals of the fission process, its development and its industrial scale deployment are now well known. This is in contrast to several of the renewable forms of low carbon electricity generation, such as solar photovoltaics and wind turbines. In many respects these technologies are in an analogous position to nuclear power in the 1960s. In considering a level playing field between technologies in the electricity generation market it would seem only fair that renewables receive substantial public subsidy at this time. Recent policy announcements in the UK and the US are consistent with such thinking.

Essential for an appreciation of the economics of nuclear power *vis-à-vis* its competitors, is an appreciation of the relevance of internalized and externalized costs. Under pressure of impending market liberalization and privatization, nuclear utilities were forced in the 1980s and 1990s to internalize the costs of plant decommissioning and radioactive waste management. Renewable electricity generation sources score well in this regard as there are no significant wastes generated, and end of life decommissioning is relatively straightforward. Importantly, however, at present all fossil-fuelled electricity generation fails to internalize the substantial environmental costs of their gaseous wastes. Prominent among such wastes is the carbon dioxide arising from hydrocarbon fuel combustion. This harmful 'greenhouse gas' is predicted to be the cause of widespread future harm to global climate (as discussed later in this chapter). Only the nuclear and renewables industries can claim to have their decommissioning and waste costs properly internalized. They are both low CO_2 electricity sources. In considering such internalization of costs, stakeholders argue as to whether or not the funds set aside for nuclear plant decommissioning and waste stream management will indeed be sufficient, but it would certainly seem that the nuclear industry has finally made an attempt to cover for such eventualities.

In chapter 5 we shall discuss the financial difficulties at the British nuclear utility, British Energy. One consequence of these difficulties was that only a few years after the establishment of the segregated fund for the decommissioning of the long-term liabilities of British Energy, the company found itself unable to make the required payments into the fund from its own resources. It seems likely that much of this burden will now fall on the British taxpayer for the long term. Electricity prices can vary greatly and therefore all predictions concerning the economic performance of nuclear utilities must be made with some caution.

An appreciation of operating costs is fundamental to an understanding of the economics of today's installed nuclear capacity. In 2001 the International Energy Agency (IEA) published a major report entitled *Nuclear Power in the OECD* [17]. Figure I.3.7 is extracted from that report (page 124) and shows operating cost breakdowns for two types of reactor technology (attributes of both these reactor technologies are discussed in detail in chapter 5). The two types considered are the Pressurized Water Reactor

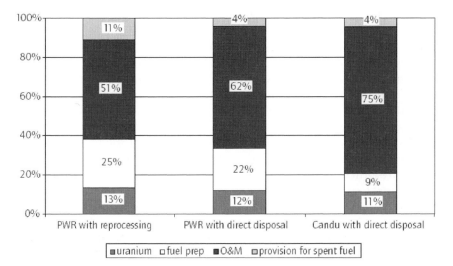

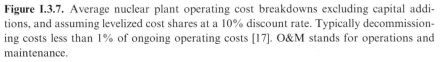

Figure I.3.7. Average nuclear plant operating cost breakdowns excluding capital additions, and assuming levelized cost shares at a 10% discount rate. Typically decommissioning costs less than 1% of ongoing operating costs [17]. O&M stands for operations and maintenance.

and the Canadian CANDU heavy water moderated reactor. Figure I.3.7 shows the proportionate contributions to the operating costs of a PWR in two scenarios (with a reprocessing-based fuel cycle and with a direct disposal once through fuel cycle). There is a matching breakdown of costs for a CANDU reactor with direct fuel disposal. It is unsurprising that the CANDU technology has lower fuel purchase and preparation costs because traditionally the CANDU system uses natural uranium fuel without enrichment. CANDU technology also matches well to the once-through fuel cycle.

As noted earlier, the operating costs of nuclear power include high levels of internalization compared with other major electricity generating technologies. In particular significant attempts have been made to internalize the environmental costs of the industry. Other factors of national energy policy that remain problematic in the design of liberalized electricity markets are reliability, security of fuel supply and generation capacity margins. Conventional nuclear power plants are well suited to continuous high quality base load electricity generation. 'Renewables' on the other hand can suffer from intermittency and poor power quality, although important progress is now being made in both these areas.

In comparing the economics of competing generation technologies it is necessary for policy makers to include properly the requirements of national energy policy if the market is to reflect national goals. One important factor that one might expect to be readily accommodated by a liberalized electricity market is the issue of technology development and licensing. Generally private sector experience in technology development and deployment exceeds that of the public sector and it would seem natural that as electricity generation becomes increasingly privatized, links to the contractors involved in the power plant construction business would become easier. However, the old paradigm of state control of electricity prices, generation and distribution in vertically integrated systems matched naturally to state control of national industrial policy. Nuclear power stations were designed and constructed by national champions who might, in time, be expected to sell or license their nuclear expertise to emerging markets. While much of the faith in the relationship between nuclear generation and national industrial capacity might now be regarded as a victory for hope over experience, that structure of state control spanning from power plant construction to electricity sales did lead to reliable, high quality electricity supplies with adequate security of supply. In addition, the countries involved in such developments have remained global leaders in science and engineering and continue to lead in all measures of technological innovation. The idea of a positive relationship between monopoly and innovation has been taken up by the *Economist* magazine as the subject of one of its recent Economics Focus columns [18]. Nuclear power may well be a good example of high levels of innovation in nationalized monopoly structures with supporting evidence of a decrease in innovation as the industry moved into a more competitive situation. It is now abundantly clear, however, that the days of innovative monopoly were characterized by high electricity prices, which acted to hamper national competitiveness across the entire economy.

The market of electricity generation and supply should in principle be able to accommodate easily issues of innovation, licensing, design improvements and new technologies. In fact, for the most part, it seems that the liberalized markets have led to electricity generation technology developments that are less innovative and fast-paced than were seen in the old days of state control. Perhaps we now are seeing the proper level of such innovation within the industry. Furthermore, it seems likely that across the economy as a whole, innovation will have increased as a result of these changes within the energy sector. It is argued here, however, that aside from the important developments in the area of renewables (where substantial state subsidies are becoming available), the liberalized markets are not yet delivering sufficient research and development to ensure the long-term needs of the United Kingdom. Notwithstanding general shortcomings, one can, however, observe a recent increase in effort devoted to new reactor designs. This is of vital importance for the possibility of any nuclear renaissance. The Generation IV International Forum is particularly important in this regard and this is discussed in detail in chapter 8.

This book does not argue for a return of the days of the welfare state for nuclear power. What is observed, however, is that the liberalized markets of North America and western Europe have, thus far, failed to capture properly all aspects of energy policy and that these shortcomings have disproportionately harmed new investment in nuclear generation capacity.

A key factor in the economics of nuclear power relates to the long lead times associated with developing new nuclear power stations. The IEA report points out (page 133) that within the OECD in the years since 1984 power plant construction has taken an average of 12 years [17]. A second distinguishing feature of nuclear power economics is the very high up-front capital costs of nuclear power plant construction and licensing. The IEA report notes (page 130) that capital costs for recently built existing nuclear power plants have tended to be roughly US $2000 per kWe installed, which compares with US $1200/kWe for coal-fired plants and only US $500/kWe for combined-cycle gas turbines. In this respect nuclear plants are reminiscent of renewables such as wind and solar photovoltaics, which also have high capital costs. For nuclear, capital costs are estimated by the IEA to be 60 to 75% of the total generating cost.

Nuclear power capital costs fall into two broad categories: the physical plant and the project finance costs. Costs associated with the physical plant include labour, materials and infrastructure, while the project finance costs include the cost of capital (discount rate over project development time), the ability to benefit from economies of scale and learning. The nuclear industry and energy policy makers recognize the difficulties caused to highly capital-intensive technologies by the move to liberalized electricity markets and significant attention has been devoted to better understanding the problem and to reducing capital costs as far as possible [20]. (See figure I.3.8.)

Clearly nuclear power, with its large up-front capital costs, is unattractive when costs of capital (discount rates) are high. Importantly the cost of such capital is completely beyond the control of the nuclear industry. In simple terms it is given by the return that the plant developers would have to pay to match the returns on offer to investors from other parts of the

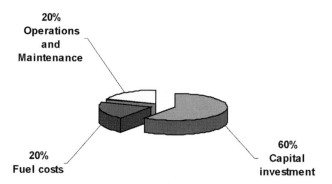

Figure I.3.8. Approximate breakdown of lifetime costs of a nuclear power plant. Capital investment is the most significant factor in the economics of nuclear power. Usual estimates place capital investment at approximately 60% of total cost [20].

economy. The IEA report notes (page 131) that doubling the discount rate from 5 to 10% increases the levelized cost of nuclear electricity by an average of 50%. In comparison, the equivalent figures are 28% for coal-fired generation and only 12% for combined cycle gas turbine generation [17].

As Malcolm Grimston and Peter Beck point out, issues of economic risk are central to nuclear power plant investment decisions [19]. There have been numerous examples of construction cost over-runs and delays. Regulatory risks are significant in nuclear power plant planning and licensing. There are also risks of generic defects in design and of accidents during construction. In a free market these risks would all fall to the investors before a single unit of electricity had been sold. Investors confronting issues of nuclear power would therefore seek to internalize these issues of economic risk via a substantial risk premium.

Investors' negative perceptions of the economic risks associated with nuclear power is in addition to any aversion to invest that might arise from political attitudes among the community of investors. For instance, early ethical investment funds typically refused to invest in any nuclear power related projects. Recently with the explosion in these forms of investment the ethical fund management community have become more pragmatic and discerning. This has been characterized as being a move from a dark green to a light green perspective on investment decisions [21]. Nevertheless with the growth of such lifestyle investments the future of nuclear power is vulnerable in ways that did not apply in the old days of state control.

Grimston and Beck's argument goes further, in that the economic risks of nuclear power tend to fall to the private investors during the ten years, or more, of power plant construction and licensing. By contrast, however, the economic risks associated with gas-fired generation tend to relate to the operational phase and the risk of primary fuel price volatility once electricity is being sold. In the case of gas-fired generation these economic risks do not fall on the original investors in the construction, but rather are passed directly to electricity consumers.

In the event that the gas-fired contribution to national electricity supply exceeds the capacity margin, some gas-fuelled electricity will inevitably be required to fulfil market demand. This condition is also true of several widely used fuels, and each such generation source (e.g. coal, gas and nuclear) should be regarded as contributing to electricity supply at all times. Depending on the design of the market, and the responsiveness of the generation type, some of these contributors are more likely to be price making and others are more likely to be price taking in the market. In the British NETA market the extent to which a generation type is price making depends largely on its flexibility. Neither nuclear power nor renewables are easily able to follow fluctuations in demand. In the previous UK electricity market, with its characteristic pool, all bidders benefited from the high prices set by the market makers, but in the NETA arrangements, with its greater emphasis on longer-term contracts, prices are more often set in bilateral trades. As a relatively inflexible base-load technology nuclear generators are forced to secure contracts to supply. Aware of nuclear power suppliers' desire for stability and long-term planning, purchasers will probably force deals at lower prices than would arise from short-term contracting. It is interesting to note that NETA is actually dominated by medium-term contracting (e.g. one year) rather than the long-term contracts originally predicted. While flexibility is indeed a virtue in national energy policy, it is by no means the only consideration: for instance, the ability to provide reliable base load (nuclear) or to address environmental policy goals (nuclear and renewables) might be regarded as being of equal, or greater, importance. It would seem, however, that flexibility of generation has a disproportionate importance in the current UK market. Those who can be flexible generally do well.

All policies have risks and the economic risks in having a market in which one flexible generation technology (gas-fired CCGT) has disproportionate power fall largely to the electricity consumer in the form of higher bills in the event that the flexible provider's primary fuel supply suffers price volatility. The more conservative technologies may cost more when the market is working well, but this comes at a severe penalty in times of market upheaval caused by what the market designers would regard as externalities, such as wars and industrial action.

The current emphasis of the English and Welsh electricity market on flexibility and short-termism seems therefore to come with a risk to the national interest. If a nuclear power plant is constructed it can reasonably be expected to operate trouble free and relatively inexpensively for 40 years or more. Long periods of profitable trouble free operations in fact allow full life-cycle assessments of the economics of nuclear power to be attractive in principle. Unfortunately, however, these timescales of nuclear power are poorly matched to the timescales of the human experience. The current time horizons of the investment community are particularly poorly matched to the time scales needed to bring new nuclear power plants on stream.

While investors in combined-cycle gas turbine power plant construction are able to frame their investment decision very closely and are able to hedge their risks, the vast majority of individual electricity consumers are not able to hedge against the risks of electricity price volatility associated with an over-reliance on gas. The consumer might eventually be able to hedge such risks through sophisticated supply contracts, but this is still many years away. Even if such hedging by consumers were possible, the fact is that nearly all consumers are, and probably always will be, very poorly informed about these issues. There is a very great imbalance of knowledge between electricity consumers and energy investors on matters affecting the economic risks of electricity generation. In particular, there is increasing concern among energy policy makers for the fuel poor and other vulnerable electricity users

Table I.3.1. Comparative electricity generating cost predictions for 2005–2010 (source: Uranium Information Centre [8], page 15).

	Nuclear	Coal	Gas
France	3.22	4.64	4.74
Russia	2.69	4.63	3.54
Japan	5.75	5.58	7.91
Korea	3.07	3.44	4.25
Spain	4.10	4.22	4.79
USA	3.33	2.48	2.33–2.71
Canada	2.47–2.96	2.92	3.00
China	2.54–3.08	3.18	—

US 1997 cents/kWh, discount rate 5% for nuclear and coal, 30 year life, 75% load factor.

[22]; but even more widely, the vast majority of consumers must be considered vulnerable as almost none are able to insulate themselves from electricity price shocks arising from volatility in the future price of natural gas.

Current markets are far from complete in regard to the construction of new capacity and the inclusion of the needs and concerns of all stakeholders. The British Government has reaffirmed that it does not believe that it should be defining the fuel mix of generation in the UK electricity market [9]. This author, for one, is not yet fully persuaded of the merits of that position.

Estimates of the current costs of nuclear electricity vary. The Uranium Information Centre in Australia quote IEA OECD data to report that nuclear costs are estimated to lie in the range 2.47 to 5.75 US cents per kWh [8]. These projected costs assume a 5% discount rate (a figure that is unrealistically low for competitive electricity markets) a 30 year plant lifetime and a 75% load factor. The nuclear projections are shown for various countries and are compared with competitor forms of generation in each case. The full results are shown in table I.3.1.

The UIC report notes that costs for several US nuclear power plants escalated dramatically during construction. Similar experiences occurred in the UK. For instance, the Dungeness B Advanced Gas-cooled Reactor took 20 years to build, not the three years originally expected [23]. As we shall see, such negative experiences are unlikely to recur in future in the event of any new build. Also, even those plants whose construction proved to be so expensive are now operating efficiently and compare favourably with coal and are cheaper than gas [8]. As the UIC report points out, 'fourteen of these older US reactors changed hands over 1998–2002 and the escalating prices indicated the favourable economics involved'.

While gas is clearly the main competitor to new nuclear build in most countries, coal should not be neglected entirely. Coal which once was extremely economically attractive for electricity generation has lost competitiveness primarily as the distances have increased between the sites of greatest electricity demand and the sites of coal mining. Increasingly cheaper coal is sourced from regions farther from the coal-fired power stations and as a consequence transport costs figure more significantly [8]. Fundamental to the future of the economic competitiveness of coal must be its gaseous wastes. Already controls on sulphur emissions favour more expensive low sulphur coals and in Europe the Large Combustion Plants Directive (2001/80/EC) will probably force much older coal generation capacity to close. In future, moves to account for the costs of carbon dioxide emissions will inevitably have a negative impact on the economics of coal.

Nuclear generation cost predictions have been prepared by the Imperial College Centre for Energy Policy and Technology (ICCEPT) for the British government. ICCEPT estimates present nuclear generating costs for newly constructed plant as being around 5.5 US cents per kWh with prospects of reductions to 4.5 ¢/kWh [24]. The differences in estimates between commentators are important because as the UIC points out:

> *A 1997 European electricity industry study compared electricity costs from nuclear, coal and gas for base-load plant commissioned in 2005. At a 5% discount rate nuclear (in France and Spain) at 3.46 cents per kWh (US) was cheaper than all but the lowest-priced gas scenario. However, at 10% discount rate nuclear, at 5.07 ¢/kWh, was more expensive than all but the high-priced gas scenario. (ECU to US cents conversion June 1997)*
>
> (Source: Uranium Information Centre [25])

Despite the weaknesses of current markets in terms of capturing all factors relevant to the decision to construct new nuclear plant, the industry has moved a long way in improving its economic attractiveness to investors. In terms of structural simplicity and lower capital costs there has been a move to the modularization of designs (easing construction) and to passive safety systems (requiring fewer components through smarter initial design). These developments are discussed in the context of various renaissance reactor designs in chapters 5 and 6.

Since the early 1990s installed nuclear power plants have greatly improved their efficiency. The May 2001 report of the US National Energy Policy Development Group chaired by the Vice-President Dick Cheney notes on pages 3–10 that for the US [26]:

> *In 2000 for the fourth year in a row the number of unscheduled reactor shutdowns was zero. The industry generated 91.1 percent of its potential maximum output, breaking its 1999 record of 88.7 percent, far better than the typical 80 percent number of ten years ago.*

In assessments of total contributions to national electricity needs, the growth in installed plant efficiency dominates over the more politically visible

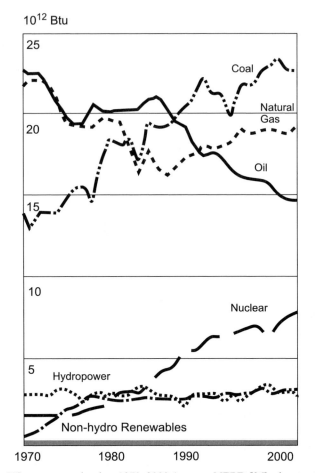

Figure I.3.9. US energy production 1970–2000 (source: NEPD [26], chapter 5).

issues of decommissioning and of stalled programmes of new build. The US NEPD (Cheney) Report illustrates this fact well with a summary of US energy production 1970–2000.

The recent rise of nuclear power in figure I.3.9 is especially remarkable when one notes that no new nuclear power plant has come on-line in the United States since the Tennessee Valley Authority's Watts Bar plant started operations on 27 May 1996 [27].

Similar growth in the nuclear power contribution to national energy needs is seen in the UK. An important part of this has been an improvement in the efficiency of the British AGR fleet, which suffered so many technical problems in the early years.

Improvements in plant availability and operating efficiency have led to reduced electricity production costs for nuclear power. In the US market

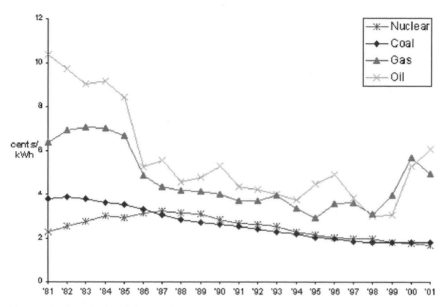

Figure I.3.10. US Electricity production costs in constant 2000 cents per kWh (source: Uranium Information Centre [8]).

since 1986 these improvements have been matched only by improvements in the performance of coal fired electricity generation. Figure I.3.10 illustrates these improvements clearly and shows costs of operations, maintenance and fuel in constant 2001 cents per kWh.

Once efficiencies of existing nuclear power plants reach close to theoretical maximum levels, it is inevitable that the nuclear power contribution to national electricity production will start to decrease as older plants are decommissioned. In order even to maintain existing levels of low CO_2 nuclear electricity generation, new build will be required. Just as operational efficiency improved with learning and operational familiarity, so it is expected that the capital costs of nuclear plant construction will also fall as experience is gained. The percentage reduction in cost arising from a doubling of production is known as the 'learning rate'. It is a fundamental issue in assessing the economics of new nuclear build. As we shall discuss, learning rates are likely to be relatively low for nuclear power technologies in the short to medium term.

In the UK the relationship between learning and nuclear power has so far largely been a story of missed opportunities. British civil nuclear power started with the Magnox programme of the 1950s and 1960s. For the most part these reactors were constructed on budget and on schedule. The second wave of British nuclear power, the Advanced Gas-cooled Reactors (AGR), were built by a range of separate contractors with poor communication between them. There was little economy of scale and very poor opportunities for learning as each plant was effectively a one-off project with more in common with traditions of craft than mass production. As a consequence, the AGR programme ran badly over-budget and massively behind schedule. The last reactor in Britain's programme of civil nuclear power was the PWR at Sizewell B in Suffolk. While based upon a US Westinghouse design, UK safety regulators and the Central Electricity Generating Board insisted upon significant design extensions and hence increased costs.

With the possible exception of the Magnox experience (which suffered from poor energy efficiency resulting from the decision to produce a power reactor technology as an evolutionary step from earlier reactor designs optimized for low temperature military plutonium production) the British nuclear industry has failed to generate the substantial learning rates that would be expected to occur in the formative days of a new technology. Today nuclear power is a mature technology and one might suppose that only relatively low learning rates should be expected. However, today's nuclear power plant designs should not be confused with plans for the medium to long term (see chapter 8). In those cases significant learning advantages could be expected. By contrast with the mature nuclear technologies of today, if properly managed, the learning opportunities for renewables should be substantial. In a debate recorded in the June 2001 issue of the magazine *Physics World* Dennis Anderson of ICCEPT reports that for renewables the learning rates on unit costs are 15–25%. However, Alan McDonald and Leo Schrattenholzer report in the journal *Energy Policy* that the learning rate for nuclear build in the OECD countries during the years 1975–1993 was only 5.8% [28]. It could be argued that this poor result for nuclear power reflects more the poor project management and the tendencies for one-off developments in that period than the actual capacity for learning in the nuclear industry if sensible deployment strategies were to be followed in the future.

For an appreciation of the economics of nuclear power, learning rates were stressed as fundamental, but difficult, in a working paper prepared for the UK Performance and Innovation Unit Energy Review (2002). As the PIU paper points out, nuclear power development in the crucial initial years of the 1950s to the 1970s suffered from completely separate secretive projects being pursued in several countries with relatively little information sharing and technology transfer. This secrecy was motivated by concerns for national security and protectionist concerns of national industrial capacity. By contrast, learning rates are expected to be highest when the units under consideration are small and relatively simple (such as with a photovoltaic cell) rather than complex and large, as in a nuclear power plant core. The PIU report also concludes that regulatory constraints are severe in the nuclear industry. All of these factors combine to cause the PIU to conclude that, looking forward, 'observed learning rates for nuclear power are likely to be quite low'.

However, while it is surely correct that learning rates in nuclear power construction have indeed been low, many of the reasons for this are now well known and could perhaps be easily avoided in future.

In most of Europe and in North America there has been little nuclear power development in the past ten years. However, to infer from this fact that there has been no learning within the industry over this period would be quite false. Western nuclear constructors such as AECL, Westinghouse, Siemens and Framatome (now combined as Framatome ANP) have continued to construct new plants and components of plants (such as Nuclear Steam Supply Systems), most notably in eastern Asia. The beneficial experiences gained in these markets over ten years or more are now available to those contemplating new nuclear build in Europe and North America.

In the following sections we shall consider various aspects of nuclear generation, all of which have links to the economics of nuclear power. In particular, in the following section we shall consider nuclear power and the environment. A fundamental issue for the relative economics of nuclear power is the prospect that charges might be levied on fossil fuel generators in proportion to the amounts of the global warming gas, carbon dioxide, that they emit.

In its report *The Future of Nuclear Power,* the Massachusetts Institute of Technology presents data on the economics of various forms of electricity generation in the face of differing levels of an emissions carbon tax. These data are shown graphically in figure I.3.11 and they illustrate that nuclear

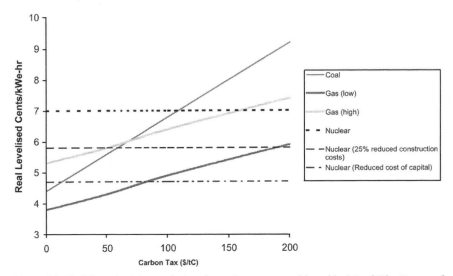

Figure I.3.11. Linearized interpolation from data presented in table 5.1 of *The Future of Nuclear Power* [29]. Graphical representation prepared by Fabien Roques, Cambridge University.

power, as previously developed, remains economically uncompetitive with gas-fuelled electricity generation at all but the highest levels of carbon taxation. However, the MIT report also considers the possibility that nuclear power can make savings in both construction costs (by reducing the construction period by one year) and in the costs of capital. If, for instance, capital cost savings were to be achieved, but construction occurred at the normal pace, then nuclear electricity is predicted to be less expensive than even low-cost gas once emissions taxes reach approximately US $80/tC. As figure I.3.11 makes clear, coal is the electricity generation technology most dramatically affected by any taxation on carbon emissions.

The UK Energy White Paper of 2003 reports that there will be no nuclear build in the UK without a new government White Paper on the subject. However, in principle one can imagine carbon charging (taxes or permits) entering the electricity market without any specific reference to nuclear power. In such a scenario, nuclear power would rapidly lose the label of being uneconomic and might plausibly be developed by private investors. Given that constitutionally government White Papers are proposals for new legislation or major policy changes, then in such a situation it would seem puzzling that a new White Paper should be needed before new nuclear build could occur. As the government itself has pointed out on numerous occasions, these days it is not the government that builds power stations.

I.3.3   Nuclear power and the atmosphere

Nuclear electricity generation has two fundamental environmental attributes. One is a significant benefit—the fact that it is a large-scale generator of electricity with almost zero carbon dioxide emissions. The second aspect has proven to be one of the greatest drawbacks of the technology and has become known as the technology's Achilles' Heel, namely, radioactive waste. In chapter 4 we shall deal in depth with issues concerning radioactive waste and plant decommissioning. For that reason only we shall not dwell upon these aspects here, but rather we shall consider nuclear power in the context of global climate change and atmospheric pollution.

While these days there is much interest in global climate change, during its period of initial deployment nuclear power was much more associated with clean smoke-free electricity that would not be a source of urban smog. Early 1950s London was largely fuelled by coal. Coal was used to heat domestic hearths in the millions of homes in the capital and coal and oil fuelled the large power stations such as Battersea and Bankside that had been built in the heart of the city in the inter-war years and early post-war years respectively. This situation led to appalling air quality problems. The history of British urban air quality is nicely summarized by the UK Environment Agency [30]:

...until the 1950s, pollution was generally accepted as the price of progress. The atmosphere of Britain's urban areas was characterized by large quantities of smoke and sulphur fumes from stacks and chimneys, which resulted in a lack of winter sunshine, pea-soup fogs, blackened buildings and black snow. No real action was taken until the infamous London smog of December 1952, which lasted for five days and was responsible for more than 4000 premature deaths. This took the form of the Clean Air Act 1956, which was later amended and extended by the Clean Air Act 1968.

The UK's first Clean Air Act and Britain's first civil nuclear power station at Calderhall in Cumbria both came into being in the same year, 1956. Politically the association of nuclear power with clean pollution-free electricity was very powerful at that time.

In the UK, however, this aspect of the benefits of nuclear power has largely faded away in recent years, probably as a consequence of the dash to relatively clean gas-fuelled generation and the concomitant decline in the use of coal for electricity generation. In addition, UK air quality has improved dramatically over recent decades resulting in public complacency.

In the United States, however, the issues of clean air and nuclear power have remained more closely coupled. So much so that compliance with the US Clean Air Acts is still a major driver in favour of nuclear power in the United States [31].

The only region of the world to have seen a major expansion in the construction of new nuclear plant in the 1990s has been eastern Asia. In China in particular there has been a significant move towards nuclear power. Among the usual energy policy drivers that any industrialized country faces, China has given particular emphasis to the relationships between coal use, national transport infrastructure, and urban air quality. The difficulties with regard to coal transport and urban pollution have given nuclear power a particular impetus in China.

The extent of the spread of nuclear power across South-East Asia is illustrated by figure I.3.12 from the Argonne National Laboratory International Nuclear Safety Center [32].

In December 1991 China's first commercial nuclear reactor (Qinshan 1) was connected to the grid. While Europe and North America were turning away from new nuclear build China continued to push its plans forward, with eight new reactors scheduled to enter service in the period 2000–2005. These reactors are Qinshan 2-A, Ling'ao 1, Qinshan 2-B, Ling'ao 2, Qinshan 3-A, Qinshan 3-B, Tianwan 1 and Tianwan 2. China has pursued this rapid process of new nuclear build in partnership with western and Russian nuclear companies [33].

In the west, where generally no such programme of new nuclear build has occurred, the main issue relating to the green credentials of

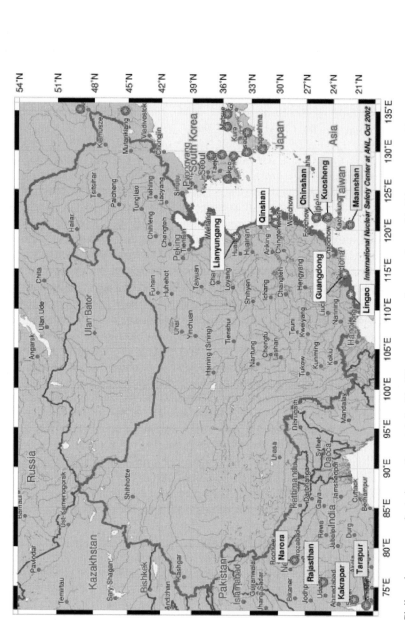

Figure I.3.12. Civil nuclear power plants in south-east Asia. The Tianwan power plants are located in Lianyungang, China (source: International Nuclear Safety Center, Argonne National Laboratory [32]).

nuclear power concerns global warming and in particular carbon dioxide emissions.

Global warming or anthropogenic climate change is a complex and at times controversial phenomenon. In essence (and somewhat simplistically) light from the sun passes through the Earth's atmosphere and warms the land and the sea. The warm surface of the planet radiates infrared heat radiation, which is invisible to the eye. Unlike the original visible light, this infrared radiation does not pass easily through some of the gases in the Earth's atmosphere (the greenhouse gases). Rather, the infrared radiation is absorbed by the atmosphere (heating it) or it is reflected back to the Earth's surface. Gases such as carbon dioxide in the atmosphere act like the glass in a greenhouse allowing the light to enter but trapping the resulting heat inside. As the atmospheric concentrations of these gases increases, so the planet warms and as a consequence the regional climates across the world change dramatically.

Carbon dioxide is the most widely discussed greenhouse gas, but it is by no means the only gas responsible for man-made climate change. Other gases of concern include methane (the major component of natural gas), nitrous oxide (a widely used anaesthetic and also a gas used to boost engine performance in sports cars), fluorocarbons (easily liquefied and volatile; these substances were widely used in aerosol propellants and refrigeration until some became implicated in atmospheric ozone depletion) and sulphur hexafluoride (an electrically insulating inert gas widely used in high voltage electrical equipment). Several factors need to be considered when comparing the harm caused by various greenhouse gas emissions.

Factors requiring consideration include the following.

- Quantity of harmful gas released to the atmosphere (often measured in millions of metric tonnes (MMT).
- Global Warming Potential (GWP)—the relative ability of the gas in question to trap heat in the atmosphere. CO_2 is the reference gas with a GWP of 1.0. Sulphur hexafluoride has a GWP over 22 000.
- Atmospheric lifetime: an indication of how long a given greenhouse gas would persist in the atmosphere. This can vary hugely from about 12 years for methane to 3200 years for sulphur hexafluoride [34].

Atmospheric lifetime differences can lead to different time integrated GWPs. For instance, the 20 year GWP for methane is 56 while the 500 year GWP is 6.5. By contrast the 20 year GWP for sulphur hexafluoride is 16 300 and the 500 year GWP for the same gas is 34 900 [34].

Despite its relatively mild GWP, carbon dioxide remains the gas of concern in the global warming debate simply as a consequence of the quantities emitted to the atmosphere each year, primarily from transport and fossil fuelled electricity production. The US Environmental Protection Agency estimates that looking ahead 100 years 82% of the global warming impact from US emissions will be attributable to carbon dioxide [34]. Despite the dominance of carbon dioxide it is important to recognize the negative impacts of the lesser known greenhouse gases, as several of these may be cost-effectively reduced in future [34].

The Kyoto Protocol is not restricted to simply carbon dioxide and thankfully the other important greenhouse gases are also covered by that agreement. The Kyoto process is, however, a complex international policy process that has yet to complete its journey. At present the United States of America has turned its back on the Kyoto mechanism. One idea originating in the United States has, however, successfully taken hold, with a widespread recognition of the benefits of emissions trading. If global carbon trading were to become a reality the resulting price of carbon could represent a major boost to the fortunes of nuclear power. Babiker and co-workers have provided a recent review of the history and progress of the Kyoto process [35].

Nuclear power inevitably leads to the release of some carbon dioxide in its fuel cycle and fabrication. In particular, it needs large amounts of cement for the physical construction of the plant (approximately one tonne of carbon dioxide is released for the production of each tonne of conventional Portland cement [36]). Despite these carbon dioxide releases, nuclear power is a negligible overall contributor to global warming. The nuclear fission process releases heat without the need for combustion and the release of the carbon dioxide characteristic of all fossil fuel electricity production. The Nuclear Energy Agency reports that nuclear power releases 2.5–2.7 g of carbon equivalent for each kWh of electricity produced. In contrast fossil fuel electricity generation releases between 105 and 366 g carbon equivalent for each kWh of electricity produced [37].

Given the negligible contribution of nuclear power to global warming one might imagine that it would be enthusiastically welcomed by the International Panel on Climate Change and the Kyoto process. In fact the relationship between nuclear power and the Kyoto process is complex, somewhat uncertain and highly political. The OECD's Nuclear Energy Agency has produced an excellent review of the relationship between nuclear power and the Kyoto Protocol [37]. The Kyoto process has four main strands. The first is that developed countries signing up to the Protocol are required to cut greenhouse gas emissions from 1990 levels by 2008–2012. An increase in nuclear power generation would be an acceptable way in which to achieve such reductions, but in reality few countries are following such an approach. A second measure is the introduction of emissions trading and, as noted elsewhere, this development could be extremely helpful to a nuclear renaissance. It is the two remaining instruments of the Kyoto process that have proved problematic for nuclear power. These measures are known as the Clean Development Mechanism and the jointly implemented projects specified by article 6 of the Protocol. In both cases nuclear power is excluded for support or subsidy on the grounds that some parties to the international negotiations

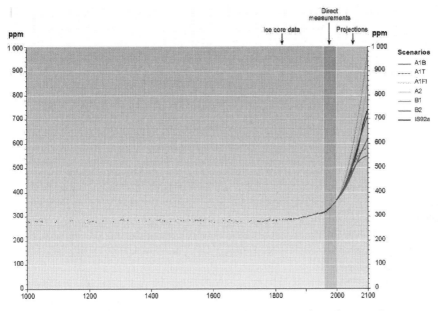

Figure I.3.13. Past and future CO_2 atmospheric concentrations (source: Intergovernmental Panel on Climate Change [38]).

felt that nuclear power is insufficiently sustainable to merit such assistance [37]. The decision to exclude nuclear power was not reached on the basis of formal and widely agreed definitions of sustainability. Nor was a careful comparison with other possible technologies made. The reasons specified for the poor sustainability of nuclear power included perceived poor safety, radioactive wastes and the risk of the proliferation of nuclear weapons. It seems that nuclear power was to be excluded for these reasons and that sustainability was a term invoked to encompass these rather diverse concerns.

Whatever the attitude of the signatories of the Kyoto Protocol to nuclear power, it is clear that it is a technology well-placed to help fix perhaps the greatest problem facing the planet. Theory tells us that mankind's greenhouse gas emissions are likely to lead to dangerous global warming. Figure I.3.13 shows the increase in atmospheric carbon dioxide concentrations during the period of global industrialization.

Consistent with the global warming hypothesis, increases in global temperature are observed. This is shown in figure I.3.14.

The forward projection shown in figure I.3.14 is especially worrying as such global temperature shifts will have profound effects including:

- sea level rise arising primarily from the thermal expansion of the oceans,
- negative impacts on global agriculture,

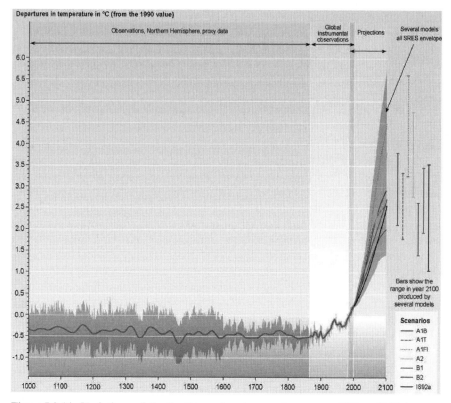

Figure I.3.14. Variations of the Earth's surface temperature: years 1000 to 2100 (source: Intergovernmental Panel on Climate Change [38]).

- worsening desertification,
- increased storms and impact on the ocean atmosphere interaction (e.g. El Niño) and
- a possible impact on ocean salinity gradients and ocean current circulation.

These issues are moving centre stage in policy discussions around the world, and yet nuclear power is only tentatively being suggested by most governments as an important piece of ammunition in the fight to restore climate stability.

I.3.4   Nuclear hydrogen

From the beginning of the development of civil nuclear energy, it has been tied to its capacity to be used for the generation of electricity. The linkages between electricity and nuclear energy are so strong that, for many, they are regarded as being two sides of the same coin. It is not, however, inevitable that the long-term path of a nuclear renaissance must lie in the same direction as developments in the electricity industry. In fact there are indications that the electricity industry is moving in directions ill-suited to sustain dramatic new developments in nuclear energy.

Worldwide the electricity industry is moving away from technocratic monopolies well-disposed to grand engineering challenges such as new nuclear power systems. Rather the electricity industry is moving towards competitive markets with very tight margins and with short-term thinking. Although for most of the 20th century all countries associated their electricity systems with their long-term industrial policy, the electricity industry has always been forced to acknowledge short-term pressures arising from the need to balance supply and demand in real time and the near impossibility of electricity storage. Improved information and communications technologies have allowed markets to operate subject to the demanding constraints arising from electricity's unusual characteristics. It seems increasingly unlikely that competitive electricity markets will be tolerant of the construction and licensing delays and the unreliability that were suffered by numerous innovative nuclear power technologies in their early years of operation.

By way of contrast the oil and petrochemical industries retain a very different culture. Their products are easily stored and this industry is expert at managing shocks and turbulence arising from a whole range of sources from the technical to the geopolitical. The culture of the oil industry is the one of exploration and risk. It has not entirely lost its affection for the speculative 'wildcatting' of the past. It is an industry that finds it natural to bet the bank on risky new ideas, including complicated engineering challenges.

Why mention the oil industry? Well, it suffers from a resource depletion problem on a timescale well matched to issues in nuclear renaissance. The oil majors are roughly one hundred years old and these have been turbulent years indeed. They are businesses that want to survive a second century and to achieve that a clear shift will be needed away from oil and towards energy services. One oil-free future for the oil majors could involve a shift towards a hydrogen economy. The idea of a hydrogen economy is one in which energy services in transport and industry are mediated by a new energy vector (hydrogen) rather than by electricity or fossil fuels [39]. Energy vectors, such as hydrogen and electricity, are not fuels themselves; rather they are ways to transport energy from one place to another. In each case the hydrogen, or the electricity, is generated from a primary source such as the burning of coal or nuclear fission.

Depending on the method of its production, hydrogen has the potential to be environmentally benign. At present hydrogen is produced commercially from the steam reforming of methane, an inexpensive process that releases large amounts of carbon dioxide greenhouse gas into the atmosphere. In the future, however, hydrogen might be produced on a large scale via emissions-free processes. One obvious approach is to use electricity (from renewable or nuclear sources) to electrolyse water. Another is the direct thermochemical cracking of water. This latter approach is particularly interesting in the context of a nuclear renaissance, as it requires very high temperatures (greater than 750 °C) [40]. In countries with high levels of bright sunshine, solar furnaces are attractive heat sources for thermochemical processing (see for instance the research under way at the Paul Scherrer Institut in Switzerland: http://solar.web.psi.ch). However, for the UK, thermochemical hydrogen production is a technology seemingly better matched to developments in high temperature gas-cooled reactors, such as those discussed in chapter 6, or even magnetically confined nuclear fusion reactors, such as those discussed in chapter 9. It is in these two high temperature nuclear technologies that the UK holds a world-leading position.

One candidate thermochemical reaction is known as the sulphur–iodine cycle. The process involves the use of sulphuric acid and iodine but, once running, the only inputs to the process are heat (at more than 750 °C) and water. The only outputs are hydrogen and oxygen [41].

Thermochemical hydrogen production using heat from nuclear fusion is especially interesting, for this form of nuclear energy is likely to be able to generate extremely high temperatures. Generally in thermochemical processes one can take the view: the hotter the better. Conventionally, nuclear fusion is expected to provide large-scale base load electricity. However, matching fusion engineering to the electricity industry needs in terms of plant size and generation reliability will be a key challenge. Great care will be needed to be sure that a nuclear fusion power station will not suffer from intermittency and break-downs during the first years of operation. It seems likely that the electricity industry will continue to be intolerant of such failures and, in the market, financial penalties for failure to supply could be very high indeed.

However, fusion for hydrogen production would seem a much more natural idea. It has potential technical benefits and business benefits for the oil majors in the middle years of this century. When compared with electricity, the hydrogen product is relatively easily stored. This could easily buffer against any intermittency and unreliability in the primary fusion heat source. Furthermore, as it is not too dissimilar to oil refining, thermochemical hydrogen production is well matched to the technical competencies of the oil and petrochemical industry. The oil majors also have an extensive infrastructure to distribute and sell fluid energy products for vehicle use. Also it so happens that the predicted resource depletion timescales for petroleum are almost ideally well matched to the predicted development time for magnetically confined fusion energy.

Fusion has the advantage over fission-based hydrogen production of higher temperatures. It does, however, have a key technical challenge—tritium. Tritium is a beta emitting radioactive gas (half-life 12.3 years [42]).

It seems likely that fusion will suffer significant difficulties in preventing tritium contamination of hydrogen product intended for sale [40]. Fortunately, as it is straightforward to check for any radioactive contamination of the commercial hydrogen product, we can be sure that there is no risk to the public. The real risk is that the thermochemical plant itself becomes contaminated with tritium, with the result that it becomes impossible to produce a hydrogen product suitable for sale.

I.3.5   Nuclear power: security of supply and reliability

In chapter 2 we considered the fundamental issues of energy policy in developed countries. Nuclear power has particular benefits and disadvantages with respect to these considerations and it is worthwhile to consider these in turn.

Regarding reliability, conventional nuclear power plants are really suited only to base load generation. Nuclear power plants can adjust their output power levels only slowly, and they are therefore widely regarded as poorly suited to load-following applications; although in France, where nuclear dominates the electricity mix, it does play a limited load-following role. Furthermore, the economics of nuclear power plants are such that the marginal cost of operations is minimal. The vast majority of the costs of the plant (capital repayment, decommissioning, safety and security etc.) occur whether or not the plant is producing any electricity. For this reason nuclear utilities are keen to operate at maximum output levels whenever they are in a position to do so. While it is true that several reactor types, such as initially the British Advanced Gas-cooled Reactors and the high temperature gas-cooled reactor at Fort St Vrain in the US have suffered from poor reliability, for the most part nuclear power plants worldwide are operating in a dependable and reliable manner. The nuclear industry would point out that the problems described above are unsurprising given that the Fort St Vrain project was technologically ahead of its time and that the British AGR programme suffered from a lack of focus and, as such, the AGR technology never fully matured. In contrast, the light water reactors and the Canadian CANDU concept succeeded in achieving technological maturity and hence high levels of operational reliability.

One aspect of the concerns for security of supply that were so dominant in the late 1970s following the oil crisis of 1973/74 is that the supply of primary fuels for electricity generation should be assured. Although the origins of these concerns relate to oil, the current topic of concern in this regard is natural gas. Coal is more widely available for utilities, and most renewables require no fuel at all. Nuclear power has as its primary fuel uranium that is imported by most countries pursuing nuclear power programmes. However, the energy density in uranium is so high, and the

fuel so easily stored, that nuclear power is regarded as having no vulnerability to fuel supply disruption. In fact, in Britain, which has no uranium
mining, nuclear fuel is regarded as domestic by energy policy makers.

Electricity quality refers to the voltage and frequency of supply. Because
of the need to control the criticality of the nuclear fission reaction carefully,
nuclear power plants operate with very high levels of stability. They are well
suited to providing very high quality electricity. As renewables increasingly
enter the generation market, care will be needed by grid managers to
ensure electricity quality, given the volatility of individual renewable
generating plants. As the proportion of renewables increases it may
become necessary for them to find ways to be able to stabilize the grid themselves. In the meantime, however, it will remain necessary for large-scale
high-quality base load generators (such as nuclear power plants) to provide
the backbone of grid supplies.

As discussed in chapter 2, generation capacity margins are increasingly
becoming a matter of concern in liberalized electricity markets. In Britain
the New Electricity Trading Arrangements (with their concomitant elimination of capacity payments for simply having unused capacity available and
ready to go) and the enforced weakening in 1998 of Innogy and Powergen's
domination of generation, led to a reduction in electricity prices and a
worrying fall in the England and Wales capacity margin to below 20% in
late 2003. More recently, prices have risen and the capacity margin has
grown somewhat, without a need for new regulatory powers. In recent
years UK electricity prices have been so low that no-one would want to
construct any form of power plant, nuclear or otherwise. It remains to be
seen whether the liberalized market in England and Wales will be able to
provide appropriate investment in generation to compensate for the rapidly
approaching closure of Britain's ageing fleet of Magnox gas-cooled nuclear
reactors. The British government continues to put much store in new renewable sources of electricity and in improved energy efficiency. While there are
some grounds for optimism surrounding new wind-turbine construction,
there is as yet little evidence of sufficient improvements in energy efficiency
to allow energy policy makers to be sanguine about UK generation adequacy.

One way out of these difficulties would be for the proposed British
Electricity Trading and Transmission Arrangements (BETTA) [9] to restore
capacity payments to utilities that maintain generating plant even if they do
not actually supply any power to the grid. Policy-makers have already started
to examine ways in which BETTA can best accommodate the growth of
intermittent renewables both technically and economically.

It has occurred to the author that nuclear plants have one attribute that
could render them well suited to providing fall-back capacity that might only
be employed in cases of national emergency, such as might occur if there were
major disruption to a primary fuel such as natural gas. The attribute is that it
would be relatively straightforward to construct a modular passive reactor

(see chapter 5) with no licence to be fuelled or to operate. This dormant plant would constitute an option for the future and allow for the rapid growth in nuclear generation should circumstances and public opinion demand it. The process for eventual licensing would need to be made clear to possible opponents and local residents but, in the British system of parliamentary democracy, parliament could accelerate that process to any extent it deemed necessary depending on prevailing circumstances. As with most insurance policies, it would be likely that this reserve nuclear capacity would never be needed, in which case the reactor would never be fuelled. In such cases the reactor would never become radioactive and would be relatively easily dismantled. The valuable resources used in the construction of the reactor could be redeployed or recycled after decommissioning. In the event that a political decision to seek permission to fuel the plant were to be made, then even an accelerated process would take some months. Thought would also need to be given to the manufacture and storage of nuclear fuel and this might need to be planned on a contingency basis. The lead-times involved would clearly not be days or weeks, but a summer of geopolitical upheaval could prompt politicians to plan to secure electricity supplies for the coming winter by proposing the use of the dormant nuclear capacity.

In power reactors that have seen service, radiological concerns have severely limited the re-use of materials after decommissioning. In some cases this policy has been rather wasteful. As experience with nuclear decommissioning grows, there is a growing body of experience with the re-use of valuable materials, such as lead shielding, once the necessary safety checks have been made.

I.3.6 Nuclear accidents and nuclear bombs

It is not the purpose of this book to dwell upon the past, and it is certainly not our intention to examine in depth outdated and obsolete technologies. However, for a proper consideration of present and future attitudes to nuclear safety it is necessary to look at some past events. Historically, expressions of public concern regarding nuclear power have been dominated by comments concerning the safety of the technology. Historically this dates back to the very earliest days of nuclear power in the 1950s. Over the years since World War II there have been clear stages in the evolving attitude of the public to nuclear power. In the 1950s and early 1960s anxieties about nuclear power were overshadowed by fears of nuclear war and nuclear weapons. Nowhere more than in the United States was the fear more palpable. Newspapers and television screens were filled with images of atmospheric nuclear weapons tests of unimaginable destructiveness and scenes of combat from the Korean War (or 'police action' as the US government preferred it to be called at the time).

The attitude of people and politicians in the US to the technologies of the Cold War (ballistic missiles, nuclear weapons etc.) were inevitably shaped by the US experience of World War II and in particular the way in which it was awakened from its isolationist slumber by the surprise attack by the Japanese Navy on its Pacific Fleet at Pearl Harbor on 7 December 1941. The US feared a sudden surprise threat to its heartland from poorly understood countries far away. It is perhaps unsurprising that in the years after World War II the United States would devote enormous resources to improved intelligence gathering and to ballistic missile based retaliatory capacity.

One key aspect of 1950s attitude to matters nuclear comes not from nuclear fission (the source of energy behind all existing nuclear power plants) but from nuclear fusion. Spencer R Weart describes on page 155 of his fascinating book *Nuclear Fear, A History of Images* [43]:

> *In short, from the outset everyone saw fusion as something that went farther even than fission bombs into the realm of the apocalypse.*

The power of the atom was clearly a frightening innovation and its growing, often subliminal, association in the public consciousness with civil nuclear power was inevitable. The timing of the awesome 1950s fusion weapon tests matched that of the early birth of nuclear fission power. The public association at a rational and an emotional level between nuclear power and nuclear weapons persists to this day, and such a linkage would inevitably be a visceral part of the public's response to any future decision to embark upon a nuclear renaissance. As we shall see, however, in the world of the early 21st century a key aspect of this concern has become the proliferation of weapons-usable materials from civilian nuclear power programmes.

Despite the pervasive fear of the 1950s and the early 1960s concerning nuclear weapons, these years were characterized by the nuclear industry's greatest ebullience and self-confidence. The media balanced their doom laden imagery of nuclear weapons tests with up-beat predictions of sparkling clean futures free of poverty and the vagaries of fuel disruption, and all made possible by cheap, reliable and ubiquitous nuclear energy. This was the time of President Eisenhower's *Atoms for Peace* speech at the United Nations (8 December 1953) in which he announced the Unites States' intention to support the establishment of an International Atomic Energy Agency [44]. Nuclear power was to be a force for good and shared widely among nations. Defence concerns were to be managed by the new IAEA through a series of inspections and visits. To this day the IAEA based in Vienna, Austria, remains a powerful force in limiting the risks of nuclear weapons proliferation and terrorist threats arising from civil nuclear power programmes.

The conscious US decision to propagandize on behalf of civil nuclear power had its clearest manifestation in a series of Atomic Energy Commission sponsored exhibitions held in cities such as Karachi, Tokyo, Cairo,

São Paolo and Tehran [45]. In 1955 the Americans even went so far as to install a working nuclear reactor complete with a characteristic blue Cerenkov radiation glow in an exhibition held in Geneva. These US initiatives ensured that in the 1950s civil nuclear power was not perceived by the overwhelming majority of people as being sinister.

One product from this period perhaps captures the attitudes of 1950s America to matters nuclear better than any other: the Ferrara Pan® candy—the *Atomic Fireball*. Popular in the US to this day this candy product aimed primarily at children has apparently not suffered from perceived negative nuclear associations. This contrasts powerfully with the 1980s experience of taking the utterly benign physics technique of nuclear magnetic resonance and bringing it into the world of medicine where it is used to image soft tissues with great effect. The medical application required a re-branding to remove the offensive and frightening word nuclear. The medical technology has now been deployed around the world under the name Magnetic Resonance Imaging, or MRI.

The 1950s and early 1960s were a unique period of technocratic optimism and public awe and fascination with matters nuclear. Importantly, however, this was not a period free of serious nuclear accidents.

Despite the public exhibitions of the Atoms for Peace era, the reality of nuclear matters was a secretive cold-war culture coupled with a continuing respect by the public for the scientists who had won World War II for the free world. These factors are likely reasons why the early accidents failed to dent public enthusiasm for nuclear technologies in ways that later accidents would.

We shall consider three serious accidents from this early period of western nuclear power. These cases are discussed in more detail by Hewitt and Collier in their book *Introduction to Nuclear Power* [14]. The first accident from this early period of nuclear optimism was the NRX incident in Chalk River, Canada. The NRX was an experimental precursor to the CANDU reactors design of Atomic Energy Canada Ltd (AECL) (see chapter 5). The NRX was a heavy-water-moderated, light-water-cooled reactor. On 12 December 1952 it suffered an emergency cooling rod insertion failure which in turn caused melting of some of the natural uranium fuel, the dissolving of fission products into the moderator and cooling water circuits and then a rupturing of these circuits that caused about one million gallons of radioactive water (containing approximately 10 000 curies of contamination) to flood the basement of the reactor. The reactor building was successfully decontaminated and experiments continued with a new reactor core installed in the same building some 14 months later.

By far the most serious of the western accidents in this first optimistic period of nuclear power was the Windscale fire of October 1957. The fire occurred in one of two plutonium producing air-cooled graphite-moderated reactors known as the Windscale Piles. The gas-cooled and graphite-moderated technology of the Windscale Piles was precursor to the UK Magnox reactor design, and even to the later programme of Advanced Gas-cooled Reactors. The Windscale fire is particularly interesting because it involved unknown physics. It represents an example of one of the public's central fears about nuclear technologies: that the scientists have failed to consider something important. The first piece of unknown physics involved in the Windscale fire was the Wigner energy (see text box, page 38). At the time that the British decided to embark on graphite-moderated reactor technology the Wigner energy was unknown to the design engineers. During a visit to the UK Eugene Wigner warned the British of the dangers of low-temperature graphite moderation. As a result of this, and their own observation of the associated swelling of the graphite moderators, the UK Atomic Energy Authority, as operators of the Windscale Piles, established a protocol by which the Wigner energy would be 'annealed' out of the graphite at regular intervals through a deliberate heating of the moderator by an operation of the reactor at a higher than usual temperature. In the course of such an annealing of Pile Number 1 it was noticed that there were parts of the moderator that retained Wigner energy. It was during a follow-up attempt to anneal those remaining zones that nuclear heating was allowed to take place too quickly, and the resulting rapid release of Wigner energy caused a fuel canister to burst. The exposed uranium fuel then ignited in the stream of air from the cooling system. Luckily the eminent physicist Sir John Cockcroft had insisted that the air coolant chimneys be fitted with filters and these captured a significant fraction of the particulate radioactivity released by the fire. As with the decision to anneal out the Wigner energy, Sir John's filters were something of an afterthought, having been fitted late in the construction process and, as a consequence, at the very top of the chimneys. It would have been far easier to design filters at the base of the chimneys, but the urgent drive for British plutonium for nuclear weapons led to such changes being incorporated during construction.

The second aspect of unknown science associated with the Windscale Piles was the way in which the fire was eventually brought under control. The reactors were designed to be air-cooled. Attempts to cool the reactor and extinguish the fire with carbon dioxide had failed during the first night of the accident. The following morning it was decided to pump water into the reactor despite high levels of uncertainty among those present as to the reactor response to this tactic. These reactors had never been designed to include water in any circumstances. Some present feared that adding water might make things far worse and even cause an explosion. Despite this, the group decided to take the risk and in fact the idea worked and the fire was extinguished safely. Despite that success, more than 20 000 curies of volatile iodine-131 were released into the atmosphere and it has been estimated that approximately 30 additional cancer deaths may have occurred in the general public as a result of the accident. This compares with the natural occurrence of 1 million cancer deaths in that same population in the same period [14].

The third major incident from the early period of western nuclear power production was the SL-1 reactor incident of January 1961 at the National Reactor Testing Station in Idaho, United States. The accident was an unforeseen criticality and power surge caused by the manual lifting of the control rods during maintenance. Three reactor operators were killed and the metal plate fuel melted and fell into the cooling water where it generated very large and sudden steam pressures causing the reactor vessel to be thrown three metres into the air, before crashing back to its original position.

The three major early nuclear accidents briefly discussed here illustrate respectively the dangers of equipment failure, unforeseen science and operator error in reactor safety. These stories from the 1950s and early 1960s had sufficient gravity to alert the public to serious safety concerns with nuclear power. Yet, as we have discussed, there was little public concern at the time.

The real opposition to nuclear power within the public grew in the 1970s and the 1980s. It may be argued that this has been a consequence of the rise of single-issue pressure groups and youth culture. That is, as the anti-Vietnam War demonstrations of the late 1960s grew out of earlier Civil Rights demonstrations, so the anti-nuclear demonstrations of the late 1970s arose directly from the Vietnam War protests, once that conflict had come to an end. This, however, is a rather Americanized perspective on what has been an erosion of enthusiasm for nuclear power. In Britain the defining socio-political events of relevance are those associated with the rise of the Campaign for Nuclear Disarmament (CND) in the late 1960s and resurgently in the early 1980s. Not only was CND passionate and anti-American, but it was also fun and it was cool. This fusion of popular culture with the British anti-nuclear movement of the 1960s is vividly captured by the present writer's uncle Jeff Nuttall in his visceral autobiography *Bomb Culture* in which he describes one CND Aldermaston march as a Carnival of Optimism [47]: 'Protest was associated with festivity'. This important aspect of matters nuclear has only slightly attenuated with the passing decades. Those advocating nuclear renaissance ignore such aspects of the politics of nuclear power at their peril.

The 1970s were periods of increased government openness, particularly in the United States. The US Freedom of Information Act had been signed into law by a less than enthusiastic President Johnson on 4 July 1966 [48].

In the 1970s Hollywood found a radical voice. Energized by new graduates from the University of Southern California and inspired by progressive and radical European cinema, the United States produced in a few years a series of cathartic and expressive movies on a range of important topics from the Vietnam War to the alienation of youth.

Nuclear power could not avoid such a searchlight and on 16 March 1979 *The China Syndrome* opened in American cinemas. With an all-star cast the movie deals with a fictional pressurized water reactor accident in southern

California and the attempts by the utility and others to cover up the story. Astonishingly, less than two weeks after the release of the film, reality seemed to be mirroring art with the serious accident at the Three Mile Island PWR near Harrisburg, Pennsylvania. In the early morning of 28 March 1979 a pump in the turbine hall stopped. Its function was to return water from the condensers back to reactor number 2. This automatically tripped the main steam generator feedwater pumps, which in turn caused the turbines to trip [49]. Although the reactor coolant circuit (see chapter 5) continued to operate after the trip, the stopping of the steam generator circuit meant that heat was not being removed fast enough, so the reactor automatically shut itself down by inserting the control rods. All this happened in a little over ten seconds.

Importantly stopping the nuclear reaction was not the end of the problem. It was a failure of systems designed to cope with the decay heat that led to the melting of 30–40% of the fuel in the core. A crucial valve had been left open and the warning light that could have alerted the operators was covered by a maintenance label. If this had been spotted in the first six hours of the accident the worst problems would probably have been avoided. As it was, the core was allowed to become uncovered (substantial parts became dry and completely uncooled) and large amounts of steam were first vented into the containment and then via a sump pump into the auxiliary building and from there up a chimney into the air outside the plant.

Although by far the most serious reactor accident in North America, two aspects of the reactor design helped mitigate the scale of the disaster. The first was that the steel of the pressure vessel proved to be more resistant to molten fuel and debris than had been feared and, second, the philosophy of reactor containment greatly reduced the amount of radioactivity released to the environment. The Three Mile Island disaster has ensured, beyond all doubt, that any future reactors constructed in the United States will have a true containment, whatever their designers might say is actually necessary.

As can be seen from figure I.3.15 the Three Mile Island accident significantly dented US enthusiasm for nuclear power. During this period, however, nuclear power remained a national concern (energy security, national industrial policy etc.). There were some issues of local concern, particularly in the increasingly problematic area of nuclear waste management. However, it was not until the far more serious Chernobyl accident of April 1986 that the public and policy-makers really understood the international reach of the risks associated with nuclear power. Local health and economic security were vulnerable to accidents in reactors hundreds of miles away and completely beyond the reach of local regulators and those who had been providing well-meaning assurances about the safety of nuclear power.

On 26 April 1986 the operators at the number 4 unit of the Chernobyl nuclear power plant complex were conducting an experiment to see if the

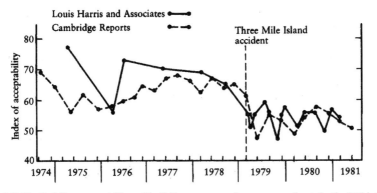

Figure I.3.15. Public acceptability of building more nuclear power plants in the USA (from *Nuclear Energy and the Public*, Joop van der Pligt [50]).

spinning turbine of the reactor would provide enough electricity during coast-down to run vital components for a few seconds. It was thought that this would add a valuable extra bit of information for the reactor safety case [51]. Unfortunately in order to conduct the experiment several of the reactor main safety systems had to be disconnected, and during the tests the operators even went beyond the parameters of the planned experiment and violated a whole series of safety requirements that were well known to them. Over a period of roughly eleven hours the operators pushed forward with their reckless experiment until at 01:24 h on 28 April the reactor building was torn apart by two massive explosions in quick succession [52].

The precise cause of the explosions may never be known, but it seems likely that the cause was far closer to being a nuclear explosion than any in the industry would have previously considered possible. That is, a major part of the sudden release of energy arose from energy liberated in an uncontrolled fission reaction arising from a prompt criticality. Hewitt and Collier suggest that the fission heat melted the fuel and the rapidly increasing fission gas pressures ruptured the fuel rods and channels and steam surged into the graphite moderator. Water then rushed back, where it combined with the fuel and there followed an explosive fuel–coolant interaction. This may have been the source of the first detonation, while the second would be a more straightforward chemical explosion of hot hydrogen gas and carbon monoxide mixed with air.

The force of the explosions completely destroyed the reactor vessel, blowing its enormously heavy (more than 1000 tonnes) top and bottom lids clean off. It is estimated that 3–3.5% of the fission product inventory of the mature fuel load of the reactor was released in the explosion and the graphite fire that followed [53]. Thirty-one people died in the explosion and its immediate aftermath. A further 30 000 cancer fatalities have been estimated to occur over the next 40 years in Russia and western Europe, although these will be hard to discern, even in aggregate, over the statistical randomness in the many millions of cancers expected in that population in the same period.

The Chernobyl cancer legacy arises more than anything from the plume of radioactive fission products that spread north-west from the Ukraine in the days after the accident. Significant radioactive contamination was deposited on highly populated areas such as Munich, Germany, in rain showers from the plume.

While the lack of a safety culture in the Soviet Union can rightly be blamed for the accident, the authoritarian command and control nature of the Soviet system may have helped the post-disaster response, albeit with huge human costs associated with those called in to manage the problem.

As figure I.3.16 illustrates, the Chernobyl disaster boosted public opposition to nuclear power in many countries. The initial public responses were more hostile than those measured one year after the accident, but nevertheless public discomfort with nuclear power increased markedly as a result of the Chernobyl experience.

As Hewitt and Collier point out [55], the Chernobyl disaster was remarkable in that at no point was there a failure of the equipment. Furthermore there was only one operator mistake, in failing to reset a vital piece of equipment needed to regulate the reactor. All other contributing factors involved were deliberate acts by the operators intent on pursuing their experiment. Furthermore, and importantly, there was no malice involved and at no point did the operators intend to do anything dangerous.

It is interesting to note the severity of an accident that can occur without any malevolent intent. Increasingly, however, policy-makers are having to confront the possibility of deliberate sabotage to a nuclear installation. One set of events more than any other has driven such a change of outlook. Those events are, of course, the terrorist attacks on the United States on 11 September 2001. It is noteworthy that those attacks were entirely separate from the nuclear fuel cycle, but their consequences have been particularly profound and far-reaching for the nuclear industry. The attacks forced those concerned with nuclear infrastructure security to cross a Rubicon and to plan not only for most likely threats, but for maximum conceivable threats.

Some have wondered whether nuclear power plants are indeed attractive targets for terrorist attack by the likes of Al-Qaeda. There seems to be little room for optimism, however. One of the attributes of the assault on the Twin Towers in New York City was that, to a western audience at least, this was an attack on a familiar part of our normal experience—the city centre sky-scraper and an international financial centre. To the alienated and impover-ished youth of certain parts of the Arab world, however, the World Trade Center towers must have been an iconic, perhaps almost abstract, image, completely divorced from their daily experience. With such an attack it

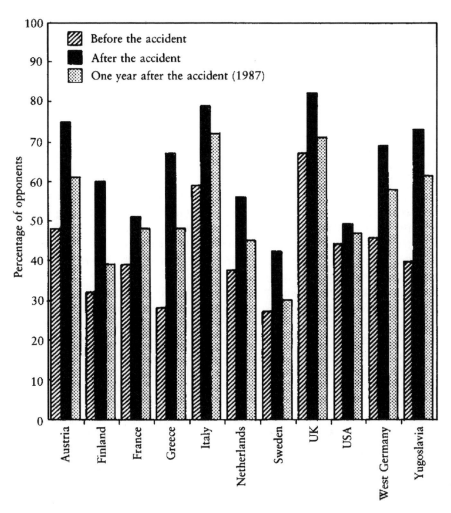

Figure I.3.16. Effect of the Chernobyl accident on the extent of public opposition to nuclear power in various countries (from *Nuclear Energy and the Public*, Joop van der Pligt [54]).

might be possible for the terrorists to secure a victory without risking an overwhelming response of sympathy for the victims from their supporters. An extension of such reasoning might allow one to infer that water supplies, which are so vital to poorer populations, would make unlikely targets for Al-Qaeda, despite the horrendous harm that might be caused from such a terrorist attack. An assault intended to destroy a nuclear power plant, on the other hand, would (as with the WTC attacks) be something that would profoundly shake the populations of the developed world, but generate little empathy with the victims from within the terrorist heartland.

The War on Terror has also caused a shift in thinking associated with one of the longest standing concerns of civil nuclear power development—nuclear proliferation. During the Cold War nuclear proliferation was centred on the concern that individual nation states might develop the ability to produce their own fission weapons and hence destabilize the delicately balanced international order. Since the Cold War this concern has continued at several levels.

The presence of nuclear materials in states descending into turmoil such as the former Yugoslavia or the former Soviet republic of Georgia, which transferred its research stock of 5 kg of highly enriched uranium (HEU) to the United Kingdom in 1998 [56].

Nuclear weapons have been developed independently by several countries, such as Israel, South Africa (which unilaterally destroyed its small stockpile), Pakistan and India. Since 11 September 2001 and President George W Bush's Axis of Evil State of the Union speech of January 2002, attention on state-sponsored proliferation has centred on Iraq (with the later realization that no weapons of mass destruction were to be found), Iran, Libya and perhaps most challengingly of all North Korea.

Concerns of homeland security have, however, broadened from the original Cold War concern that a nation state might launch a well-engineered nuclear weapon on a western nation state, to a fear that a sub-national body, such as a terrorist cell, might detonate an improvised low yield fission weapon, or even more likely a radiological weapon or 'dirty bomb' in an urban area.

Those involved in planning for a nuclear renaissance must now recognize the increased need to protect nuclear infrastructures. Furthermore, they must accommodate risks that nuclear materials might be diverted by skilful well-resourced terrorists intent on wreaking maximum harm with little or no regard to their own personal safety. A terrorist weapon improvised from materials obtained from the civil nuclear fuel cycle (i.e. based upon plutonium, not highly enriched uranium, which has almost no association with civil nuclear power), would be most unlikely to resemble nuclear weapons as commonly conceived. If the terrorists were to attempt the ambitious task of constructing a plutonium fission weapon then they would be most unlikely to achieve the level of explosive yield commonly associated with atomic bombs. In fact they would probably fail to achieve any fission yield at all. They might also attempt to achieve an uncontrolled thermal criticality of fissile materials or to construct a radiological dispersal device (a 'dirty bomb'). In such cases the physical damage would be far smaller than from even a low yield fission explosion. In the absence of a fission explosion, probably the greatest destructive capacity of all these actions would be the vast amounts of fear and panic created. The experience would likely be utterly devastating to those caught up in the crisis. Once the panic had subsided, it would also be likely that the civil nuclear industry would have suffered a serious blow in the public's perceptions, perhaps even a mortal blow. Those planning for a nuclear renaissance must continue to be sensitive to all aspects of this complex problem.

I.3.7 Nuclear power industry and skills

Of all the factors that shape the future of a nuclear renaissance in western Europe and North America only one has the status of potential 'show-stopper'. This factor is not the safety of nuclear power, its environmental consequences or its economics: it is the supply of engineers, scientists and skilled trades needed to keep the industry operating and properly regulated. The age profile of the industry is dangerously skewed towards those facing retirement, and not to those seeking new challenges. The skills shortage is not just an obstacle to new build. It is also a concern for the future of the nuclear power industry in all scenarios, including phase-out and decommissioning.

Internationally the leading forums for these issues have been the Organisation for Economic Co-operation and Development's Nuclear Energy Agency and the International Atomic Energy Authority. The OECD report of 2000 urged governments to tackle issues in nuclear education and training as a priority [57].

In the UK the Health and Safety Executive has been monitoring the issue of nuclear education in British universities since 2000. In March 2002 the HSE published its second report on the issue and strikingly led with the primary conclusion that '*if nuclear education were a patient in a hospital it would be in intensive care*' [58]. The HSE recommends that:

> *The nuclear industry needs to identify what competencies it is going to need in the short-term, as well as the medium- to long-term, and work with universities to ensure that courses are in place to meet those needs. That in-house training can be pursued in the absence of an external knowledge base is a non sequitur.*

The clear concern of the HSE at the limits of the industry to address its own needs runs counter in tone to an observation from the British Government's Nuclear Skills Group that reported in December 2002 [59]. The Nuclear Skills Group state that:

> *A common view among employers is that they need generalist engineers and physical scientists who can be given specialist in-house training in nuclear technology. As a consequence, there is low demand for specialist nuclear education in Higher Education Institutions.*

There is clearly something of a difference in perception. If the HSE is correct, it implies that the UK nuclear industry is wrong when it reports that there is little need or demand for nuclear specialists. Clearly, the NSG report is less vehement in its pessimism than its HSE precursor, but this perhaps reflects the NSG's own observation that within the nuclear power industry the problematic hot spots are safety case production and radiological protection—two core considerations of a nuclear safety regulator. While the HSE study was not tasked with assessing the regulator's own requirements, it is perhaps natural that the regulator might have a particularly pessimistic opinion on these matters. The NSG, however are not wholly sanguine in their remarks. The NSG notes that for graduates:

> *15 500 graduates required by the power fuel, defence and clean up sub-sectors over the next 15 years equates to approximately 1000 graduates per year. Of these, 700 are replacements for retirements and 300 are a response to growth of nuclear clean up. By comparison the sector's 2001 graduate recruitment target was approximately 560.*

The current shortfall in UK nuclear graduate recruitment identified by the NSG arises under circumstances of no nuclear renaissance. Any decision to embark upon new build would need to be fully cognisant of the limited human capacity of the industry to fulfil its existing obligations.

A particular concern of the UK in the area of nuclear skills relates to skilled trades. The NSG indicative scenario that gave rise to the graduate recruitment estimates quoted above also suggests that the UK will need to recruit 7850 skilled trades people into the industry over the 15 year time frame of the scenario. This is a particular challenge given the erosion of apprenticeship training in such skills in the UK in the 1970s and 1980s [60].

The labour market for highly qualified workers is increasingly globalized. This implies that not only would a local decision to embark on nuclear new build jeopardize the weak local skill base required for pre-existing missions, but a decision by a competitor nation to embark on new build could jeopardize key parts of the local skills base. A particular example might be that a firm US resolve to embark on a nuclear renaissance might lead the US to recruit nuclear engineers from other countries, such as the UK. As noted above, this might jeopardize UK capacity to meet its existing nuclear skills needs for decommissioning, and it may also remove the ability to keep the nuclear option open and thereby prevent any UK nuclear renaissance. Depending on the extent of such international skills flows, it is possible to imagine a first mover advantage among the larger players.

In positing the idea of a global competition for talent in nuclear engineering and allied professions, we may have failed to attend to the fact that the nuclear industry is not a typical globalized industry. The relationship between nuclear skills and nuclear weapons proliferation is important and would block some flows that might otherwise occur. The primary competition between the G8 countries is, however, very real and most probably a real threat to the capacity for nuclear renaissance in some smaller countries, such as the UK, that might procrastinate on nuclear energy policy for too long.

So how did countries with significant existing nuclear infrastructure and with strong science, engineering and technology backgrounds find themselves in this situation? In the context of health physics, the UK NSG notes that career path norms can encourage workers to gradually lose their technical skills as they move into management. This reality is something that policy makers must accept. Any discipline in which management opportunities are reserved for those with backgrounds outside the field will find it difficult to retain ambitious young people who had initially been recruited into technical positions. It is better to see such people progress into management positions within the industry than to see them leave the sector entirely.

According to the NSG, potential recruits find the apparent indecision of the industry most off-putting. While policy studies demonstrate that the nuclear industry will inevitably be a rich source of employment opportunities in the future, the uncertainty concerning the trajectory of the industry (e.g. decommissioning versus new build) and the uncertain time-scales of next steps (long consultation periods etc.) cause the industry to be unattractive to new recruits.

In an era of increasing career mobility the nuclear industry suffers from the fact that its highly qualified workers possess highly transferable skills. Other sectors value workers' experience of the nuclear industry safety culture and engineering systems approaches. While other industries can easily draw upon the pool of talented nuclear workers it is extremely expensive for the nuclear industry to draw in talent from rival fields. This is because nuclear workers require high levels of specialist training unavailable in other sectors.

The NSG points out that moves towards contractorization of the British nuclear industry have led to a complex web of supplier–client relationships with no clear responsibility for ensuring the long-term sustainability of the nuclear skills base. In fact the fragmentation of the industry has gone so far that some have come to question the very existence of a nuclear industry. Such commentators suggest that there are large engineering companies for which nuclear contracts are a small part of the commercial reality and a set of smaller niche players each pursuing specialist tasks with little in the way of common attributes and responsibilities.

Finally it is important to note the NSG observation that the nuclear industry is among the most opaque and secretive sectors of employment. The secrecy and the perception of a closed community relates in large measure to the early origins of the nuclear industry in highly secretive defence work dedicated to nuclear weapons development and submarine propulsion reactors. Much of the opacity of the field relates to its particular jargon and extensive use of acronyms, which the NSG rightly points out can be alienating to those with little technical experience.

Despite the recruitment and retention difficulties faced by the nuclear industry, there is one aspect of its personnel traditions that offers some comfort to those concerned with high level nuclear skills. When compared with other contemporary high-technology sectors the nuclear industry has an unusually strong tradition of training up to senior technical positions those who enter the industry with relatively modest academic qualifications. It is for this reason that recent concerns as to the health of skilled trades apprenticeships in the industry are particularly worrying as they not only endanger the technician and operator skills base, but they could also damage the long-term health of the nuclear industry's graduate skills base, given the industry tradition to train from within.

I.3.8   Concluding thoughts

Malcolm Grimston argues that alongside all the widely discussed obstacles facing nuclear new build (such as economics, the waste question, proliferation fears and safety concerns) there is the special problem that nuclear power is decreasingly well matched to the structure of our society. Nuclear power related well to those long-gone days following World War II, when society and industry trusted that the state would be able to provide for their needs via centrally planned large-scale infrastructures such as the electricity grid. The technocrats of Whitehall would look after us and make the trains run on time. Today, however, the *zeitgeist* is very different. We live in an increasingly fragmented society with a much stronger sense of self-reliance at all levels—so much so that I might argue that whatever the fundamental economic, safety and environmental attributes of a domestic micro-turbine or Stirling engine for domestic heat and electricity co-generation, such a machine will prove more popular with householders than nuclear electricity. It seems that these days most householders place a premium on the ability to command and control their own lives as far as possible.

Chapter 4

Nuclear waste management

I.4.1   Introduction

The problem of radioactive wastes actually predates the nuclear industry. This is because of the extensive use of radioactive materials in the years before World War II for luminous paints (for clock faces and aircraft instruments), medicine (radium therapies) and industrial processes (uranium ceramic glazes, coloured glasses and lamp mantles). Interestingly, in the UK, materials from these periods remain outside the provisions of radioactive waste legislation. Effectively, therefore, the British legal clock for radioactive waste management started in 1946. We shall follow this precedent and restrict our concerns here to materials generated since World War II. We shall pay particular attention to wastes arising from nuclear power programmes in this period. We shall also refer to military nuclear issues in so far as they relate to civil nuclear policy.

The controversy surrounding radioactive waste management and the political sensitivity of the subject have led to certain simplifying axioms. Policy makers introduced these axioms in the hope that they would permit more rapid progress in this particularly tricky policy area. Examples of such axioms include the following.

- The separation of military and civilian wastes. In most nuclear weapons states there is a strong separation made between military and civilian materials. In Britain and the United States this extends beyond nuclear weapons to include a separation between the civil power fuel cycle and the fuel cycle associated with naval propulsion systems.
- A distinction is frequently made between legacy wastes and future waste arisings. This is done so as to accommodate the concerns of those resolutely opposed to any nuclear renaissance. It is believed that by restricting debate to legacy wastes a wider range of stakeholders will be likely to respond positively to invitations to contribute to the policy-making process.

- It is widely believed that there is a need to firmly classify materials as wastes or assets. Once this distinction is finalized, those concerned with nuclear waste management would then need to have little concern for assets.
- It is suggested that the principle that the polluter pays should apply to radioactive wastes so as properly and most appropriately to internalize the costs of waste management.
- The principle of intergenerational equity is often invoked as being important to radioactive waste policy. That is, those responsible for nuclear policy today should avoid handing to future generations a situation worse than the one we inherited. This is often extended to yield the recommendation that we must implement solutions now and avoid passing key decisions to those in the future.

Each of these axioms has been introduced into nuclear policy with the intention of facilitating greater progress. In this author's opinion, however, each axiom brings with it the risk that it will excessively constrain an evaluation of best options in one of the trickiest technology and policy sectors. By way of examples:

- The US Waste Isolation Pilot Plant (WIPP) in New Mexico as a facility for defence wastes is entirely separate from developments at Yucca Mountain, Nevada, associated with wastes from the civilian nuclear fuel cycle.
- Forcing a distinction between legacy wastes and future waste arisings militates against the best assessment of future nuclear policy regarding future nuclear power. The synergies between issues surrounding nuclear renaissance and nuclear waste management are sufficiently strong that any attempt to frame policy for one independent of policy for the other, runs the risk of producing sub-optimal policies for both issues. This will be discussed further in the context of nuclear waste burners in chapter 7.

    Later in this chapter we shall examine issues surrounding surplus separated civil plutonium. There have been several calls in the UK to declare this material waste, yet it clearly has the potential to be useful in some scenarios. Settling the *what is waste?* question will be certainly be slow and difficult, but there has been little discussion as to whether settling the question is truly necessary. Separated civil plutonium is clearly a problematic radioactive material requiring management, whether it is officially a 'waste' or not. As discussed in the text box (*A British Paradigm Shift?*) there can be opportunities in consciously deciding to avoid all together the tricky question *what is waste?*

- Polluter pays seems at first impression to be both fair and efficient. However, when the polluter is interpreted to be a private company there is little or no incentive for the public to endorse anything other than the most expensive solutions to the radioactive waste problem in their search for total safety. If the public were forced to pay the bills associated with their preferred strategies then they might be expected to be more modest in their requirements.

• Intergenerational equity is consistent with the principles of sustainable development. Those advocating intergenerational equity seem to do so in the hope that it will generate decisiveness and avoid procrastination. It acts in opposition to the view that concern for radioactive waste management should be a continuing responsibility. History has given us several examples of technical advancements that turned out to make matters worse than if no novel steps had been taken. For instance the widespread use of asbestos in early 20th century construction turned out to be more of drawback than a benefit, once its damaging health effects became apparent.

If a nuclear renaissance is to occur, then the industry and policy-makers must continue to increase openness and transparency in their dealings. Such openness of operations will build public trust. It should be accompanied by an opening of outlook to consider all aspects of the problem beyond the now traditional axioms. A renaissance requires not just new creativity and new build but a new receptivity to ideas and a breaking down of previous orthodoxies. In recent years, Finland has successfully adopted such an approach and, perhaps as a consequence, now finds itself in the vanguard of Europe's nuclear renaissance.

It is interesting to consider why the radioactive waste problem has been so difficult in the UK. One compelling idea is that the radioactive waste problem is an example of a wicked problem [61]. Such problems are characterized by an odd circular property that the question is shaped by the solution. As each solution is proposed it exposes new aspects of the problem. Wicked problems are not amenable to the conventional linear approaches to solving complex problems. Such linear approaches go from gathering the necessary data, through analysing the data and formulating a solution towards implementation of a final agreed solution. By contrast, wicked problems can at one moment seem to be on the verge of solution, yet the next moment the problem has to be taken back to its complete fundamentals for further progress to be made. As such, any opinion that the problem is almost solved is no indication that it actually is. Wicked problems can persist for decades and, for a true wicked problem, no solution will ever be possible. Wicked problems typically combine technical factors and social factors in complex multi-attribute trade-offs. A problem that is not wicked is said to be 'tame'. One thing is certain: in the UK, at least, radioactive waste management is not a tame problem.

I.4.1.1   Generic options for radioactive waste management

By the 1990s one approach to the long-term management of radioactive wastes had become favoured above all others—Deep Geological Disposal.

The primary intention behind this approach is to isolate harmful wastes from the biosphere for as long as dangerous levels of radioactivity persist. The basis of this isolation should be entirely passive and should not require active maintenance once the repository has been sealed. The essence is that, once disposed of, the problem has been removed and it need not concern future generations. The preferred basis to achieve such long-term isolation is the multi-barrier concept.

Each engineered barrier in the multi-barrier concept is designed to last long enough to contain a particular phase of the evolution of the wastes contained. For instance, the stainless steel drums of the Nirex concept shown in figure I.4.1 are designed to last long enough for more than 99% of the encased radioactivity to have decayed away.

Environmental groups have tended to oppose the disposal concept on the grounds that inevitably once nuclear waste is out of sight it is out of mind. A hazard for future generations is removed psychologically, but not in reality. A more cynical view can be heard within the nuclear industry. This states that because the nuclear waste problem has proven to be the most troublesome for the industry, environmental groups opposed to nuclear power will do nothing to facilitate a solution to a problem (radioactive waste management) that has provided them with so much ammunition with which to attack nuclear power as a whole. The longer the pressure groups can force

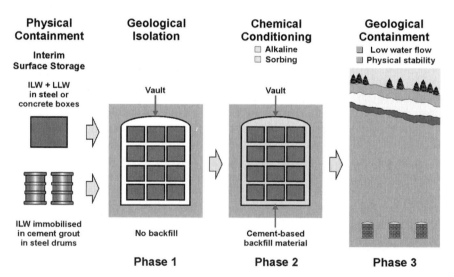

Figure I.4.1. Nirex UK multi-barrier disposal concept from the mid 1990s. The first barrier is the steel drum in which waste is held. This is designed to last at least 1000 years. The next barrier is the special cement grout or clay (backfill) packed around the drums, and the third barrier is the deep underground rock into which the disposal facility is cut (source: Nirex [62]).

procrastination on nuclear waste management then the longer they can delay any sensible assessment of nuclear power and a balanced assessment of the benefits and dis-benefits of a nuclear renaissance.

Many environmental groups, however, have made constructive suggestions for the management of nuclear waste. They tend to prefer indefinite managed surface storage. This approach represents an extension of arrangements that are currently regarded as interim in most countries. The waste should be stored in actively managed and supervised conditions on the surface. Traditionally environmentalists were opposed to burying the waste as this would be more expensive than using specially constructed buildings on the surface. The terrorist attacks of 11 September 2001 have prompted some to reconsider their position, but generally environmental groups remain implacably opposed to any form of geological 'disposal' policy for nuclear waste.

Historically other approaches have been considered. These include borehole disposal, a variant on deep geological disposal whereby backfilled boreholes are used in place of geological repositories more reminiscent of mine workings. More fancifully some have suggested that nuclear wastes should be fired into space and perhaps directly into the sun. The main objection to this approach has always been the risk of catastrophic failure on launch. Another attempt to achieve absolute disposal would be to drop nuclear waste containers into the sediment at the bottom of the deep ocean. If dropped into the right places subduction of the ocean bottom would gradually remove the waste from the biosphere. Some report that this option is still the most technically sound, but it has been completely barred by the London Convention since 1994 [63].

Finally, one other option figures increasingly in discussion of radical approaches to nuclear waste management: 'partitioning and transmutation'. As a potential technology of the nuclear renaissance, this approach is discussed in some detail in chapter 7.

I.4.2 British nuclear waste management

Each country with a radioactive waste legacy has categorized its own inventory. In the UK classifications follow naturally from the output streams of reprocessing rather than, for instance, on the basis of radioactive half-life or the chemical properties of the wastes. The official UK categories of radioactive waste are as follows.

I.4.2.1 Very low level waste (VLLW)

In recent years these wastes have been limited to beta and gamma activity less than 400 kBq per 0.1 cubic metre [66]. They are in most ways excluded from consideration as radioactive wastes. In the UK these wastes are permitted for incineration or disposal in conventional landfill. VLLW is not included in official inventories of UK radioactive waste. VLLW is relatively uncontroversial and is therefore only a minor part of nuclear waste policy and as such we shall not consider it further here.

I.4.2.2   Low level waste (LLW)

These wastes consist of potentially contaminated clothing, paper towels, plastic wrappings and metal scrap, much of which is not radioactive at all [65]. Many of these materials are treated as radioactive waste on a precautionary basis. These wastes are defined by having a specific activity higher than that of VLLW but lower than 12 million Bq/kg for beta or gamma activity and 4 million Bq/kg for alpha activity [64]. On average, given the precautionary policies adopted, the activity of the LLW destined for disposal is far lower than the upper limit. For four decades British wastes of this type have been disposed of at Drigg in Cumbria. In recent years the wastes have been compressed to reduce volume. The compacted metal containers are placed in large steel shipping containers in a shallow buried repository constructed using cut and cover techniques.

Of the three controlled categories of official nuclear waste in the UK, LLW has been the least problematic for policy-makers. The processes of collection, packaging, compression and disposal at Drigg are operating smoothly, even though the capacity of the Drigg facility is limited.

I.4.2.3   Intermediate level waste (ILW)

Internationally, the term intermediate level waste suffers from the different definitions adopted in different countries and by different agencies. In the UK the term is quite precisely defined. Most of these wastes are metals. The wastes are defined as being those with a specific activity above the range specified for LLW activity [64], but importantly for which there is no significant self-heating as a result of radioactive decay. The UK definition of ILW allows for the inclusion of some types of plutonium-contaminated wastes and these are not separately classified in the British schema. UK ILWs include the claddings stripped from spent fuel during reprocessing.

The International Atomic Energy Agency has recommended that ILW be divided into long-lived and short-lived wastes [64]. This recommendation has not been adopted and at present half-life plays no role in UK waste classification.

UK ILW is stored in shielded tanks, vaults or silos at the site where it arises [65]. Because of the reprocessing facilities at Sellafield in Cumbria large amounts of ILW are generated there. One method of treatment for ILW has it immobilized in a cement matrix and encapsulated in specially designed 500 litre steel drums. These drums are held in a surface storage facility at Sellafield, and new facilities for this purpose are under development. These stores are required as the UK lacks any long-term management strategy for these materials. ILW wastes have proved the most problematic of all for British policy-makers and this will be discussed further later in this section.

I.4.2.4 High level waste (HLW)

High level wastes are intensely radioactive. The primary defining characteristic of HLW in the UK is not actually based upon radioactivity but rather on thermal properties. HLW is defined by the attribute that it is self-heating from the radioactive decay of the waste itself (primarily of the fission products in the waste). The main radioactive components of HLW are the long-lived fission products such as technetium-99, caesium-135 and selenium-79 and minor actinides such as americium-241, neptunium-237 and curium-242. The HLW output of reprocessing is a liquid solution of the fission products and minor actinides dissolved in nitric acid. The conditioning of HLW consists of chemical reduction, evaporation, vitrification with boron (to absorb thermal neutrons) and the preparation of large sealed steel flasks containing the solid glass HLW. The flasks are placed in air-cooled channels in a special vitrified product store at Sellafield.

Responsibility for civil HLWs in the UK rests with BNFL. Nearly all these wastes arise from operation of the fuel-cycle; little or no HLW is expected to arise from the decommissioning of nuclear facilities. As described earlier, in the UK HLW is intimately connected with the highly radioactive nitric acid liquor that emerges from reprocessing. At present the UK has no policy for the long-term management of this vitrified waste, which is currently in secure storage at the BNFL Sellafield site. Back in 1998 Nirex and Department of the Environment, Transport and Regions (DETR) reported that 13% of Britain's 1800 cubic metres of HLW had at that point been vitrified [65]. Most observers expect that BNFL will complete the vitrification of the vast majority of the Britain's HLW in the years ahead. Today the UK is already reporting its HLW inventory as if it is all in final conditioned form. Once vitrified, the waste is expected to stay in store for the decades required, following substantial radioactive decay, for it no longer to be self-heating.

I.4.2.5 The scale of the UK radioactive waste problem

Whatever one's view of the future for nuclear power it is a fact that significant quantities of radioactive waste exist today which require proper

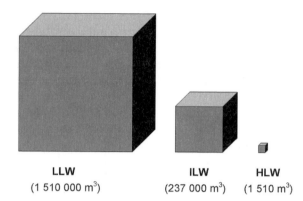

LLW
(1 510 000 m³)   **ILW**   **HLW**
                      (237 000 m³)   (1 510 m³)

Figure I.4.2. Schematic representation of the relative volumes of UK radioactive waste by official type, as specified in the 2001 UK inventory [66].

management. These existing wastes, known as 'legacy wastes', are summarized for the UK in figure I.4.2.

Figure I.4.3 highlights the relatively minor contribution that radioactive materials make to the UK's total inventory of toxic waste (note the logarithmic scale). Many chemical toxic wastes are stable for at least as long as long-lived radioactive wastes (many will be stable for infinite periods) and yet the management of chemically toxic wastes seems to cause far less

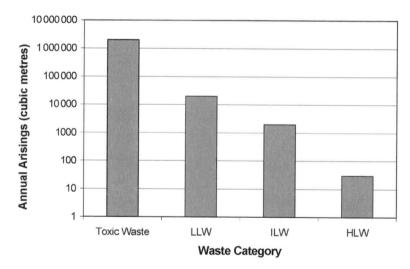

Figure I.4.3. Approximate annual wastes arising in the UK by conditioned volume. Note the logarithmic scale. Note also that the volume of conventional industrial toxic wastes far exceeds that of radioactive wastes. (Data from Jeremy Western's contribution to the 1999 UK-CEED consensus conference [67].)

public anxiety than radioactive waste. In part this may be ascribed to the 'dread risks' of nuclear power discussed by Spencer R Weart in his book *Nuclear Fear: A History of Images* [43]. Despite the existence of greater risks, the acute anxieties of the public concerning nuclear wastes are both real and probably permanent. While the extent of actual hazards is debatable, public anxieties are a definite reality that policy makers must recognize and accommodate. Nuclear policy-makers must better appreciate that the best route to the minimization of public fears is not likely to be via the single-minded minimization of danger, but rather from a full consideration of the causes of public anxiety and a sincere desire to reduce such discomfort.

The UK-CEED consensus conference in 1999 illustrated well that informed members of the British public regard partitioning and transmutation (P&T) as an important scientific contribution to the solution of a key public policy problem that they regard as having been created by science and technology [67]. This public interest and enthusiasm for P&T is mirrored in parliament where the House of Commons Science and Technology Select Committee has recommended that 'the Government monitor technological developments in transmutation and keep it under review as part of its radioactive waste management strategy' [68]. The issues are considered further in chapter 7.

I.4.2.6   The nature of the hazard

Table I.4.1 illustrates how the radioactivity contained within the wastes is almost in inverse proportion to the volumes of waste involved. This table makes clear that the most problematic wastes of all are the HLWs, while the LLWs are indeed relatively innocuous.

While British nuclear wastes are not classified by half-life, estimates have been made of the behaviour of the waste radioactivity with time (see figure I.4.4). It is interesting to note that the radioactivity of both HLW and LLW is expected to increase over the first few decades. Fundamentally this is due to predicted increases in the total amounts of these wastes as further waste is created by the reprocessing of spent fuel or by decommissioning. Another factor, but probably a negligible one, is that several of the isotopes in HLW, such as the **actinides**, have decay chains in which the

Table I.4.1. Radioactivities in TBq of UK radioactive waste stocks on 1 April 2001 (1 TBq = 10^{12} radioactive distintegrations per second) [66].

HLW	ILW	LLW
57 800 000	5 290 000	11

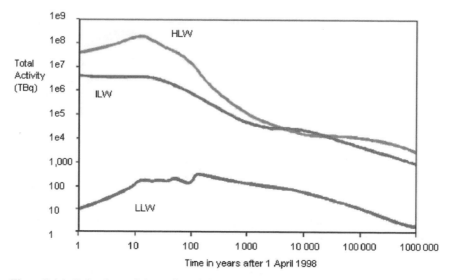

Figure I.4.4. Behaviour of the radioactivity of UK wastes with time [65].

half-lives of the decay products (which are themselves radioactive) are significantly shorter than those of the original actinide isotope in the HLW. Once the radioactive decays have finished cascading through the various actinide decay chains to reach stable isotopes, then overall HLW activity will clearly be seen to decrease. Such an effect is known as 'ingrowth'. Storing HLW for 50 years or more is sensible, as both total activity and the level of self-heating decrease significantly, thus allowing for the easier disposal of the wastes at that stage. It may even be possible formally to reclassify some cooled down HLW as ILW and hence render its management more straightforward.

I.4.2.7   Key elements from the history of UK policy

Fundamental to the history of British civilian nuclear waste management has been British early adoption of reprocessing-based fuel cycles. This in turn arose directly from the UK's decisions in the late 1940s concerning nuclear weapons. As mentioned earlier, British independent efforts to develop plutonium-based nuclear weapons following the passing of the US McMahon Act led to significant experience with both graphite-moderated reactors (the Windscale Piles) and with the chemical techniques needed for plutonium extraction from spent fuel. Following the experience gained with the Windscale Piles, the first wave of UK civilian power reactors (the Magnox reactors) were also gas-cooled and graphite-moderated and they took their name from the magnesium (no oxidation) alloy cladding to the natural uranium nuclear fuel. The magnesium alloy was needed because it had extremely low neutron absorption as required by a natural uranium-fuelled reactor. The use of Magnox fuel led directly to reprocessing as the fuel claddings were believed to be unstable in damp air and to have only limited stability when stored underwater. Conventionally Magnox fuel handling was to be done under water, but by the time of the last Magnox station, Wylfa in North Wales, an interim dry fuel handling and storage process had indeed been developed for Magnox fuel, but only that one power station was suitably equipped. It is from these early days of reprocessing in the late 1950s and early 1960s that Britain's key distinctions between spent fuel, separated civil plutonium, reprocessed uranium and ILW and HLW first arose. The fuel for the second and third waves of UK civil power reactor (the AGRs and the Sizewell B PWR) are dry storable indefinitely and reprocessing is not an essentially management strategy for these spent fuels. In the 1970s with AGR spent fuel management issues looming and with uranium prices at an all time high (see figure I.4.5), Britain faced a decision whether to adopt a reprocessing based fuel cycle for spent AGR fuel or to dry store and then dispose of the spent fuel directly as a waste.

Following an extensive public inquiry chaired by Justice Parker and his 1978 report [72], the UK resolved to construct a new reprocessing facility for oxide fuels such as those used in the AGR plants and in Light Water Reactors. While issues of environment, proliferation and energy security were discussed, the dominant considerations at the time were economic. Even with (as it turned out erroneous) predictions of continuing high prices for uranium, the economic model underpinning the UK Thermal Oxide Reprocessing Plant (THORP) required a very large scale facility. Economies of scale were believed to be the route to a profitable reprocessing business.

From the beginning it was clear that the only way in which the new facility could generate the necessary quantities of reprocessing business would be seek international contracts. BNFL as the developers of the THORP facility were assiduous in the 1980s in locking international utilities (particularly from Germany and Japan) into long-term reprocessing contracts. The sums involved were very large, as Michael Spicer noted in the House of Commons debate on the Atomic Energy Bill 1989 [73]:

> ... *contracts with Japan, which amount to around £2 billion and will make THORP the largest earner of yen in the country.*

With its development starting in the late 1970s and having been constructed at a capital cost of around £2.8 billion, THORP started its first test runs in 1994. By the mid 1990s with continued low uranium prices, it had become clear that it would be uneconomic. Moreover the separation of civil plutonium was no longer clearly beneficial following the cancellation of the UK fast reactor programme in 1994. As the British Energy submission to the 1992 PIU Energy Review makes clear, any nuclear renaissance is now unlikely to be based on reprocessing, but rather on the once-through fuel cycle long advocated by the United States for civil nuclear power [74].

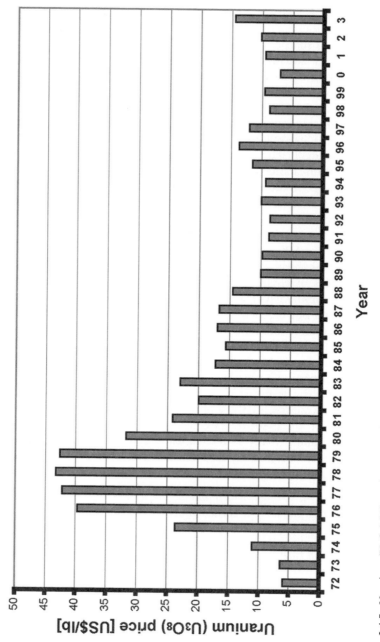

Figure I.4.5. Uranium (U_3O_8) US market spot price on an annual basis (source: World Nuclear Association [71]).

Nirex is responsible for UK research, development and demonstration for the deep geological disposal of intermediate level wastes and research into the management of some low level wastes. It was founded in 1982 and established as a company in 1985. It is owned and funded by the nuclear industry with a special share held by the British Government. In addition to its UK responsibilities, Nirex retains a watching brief on radioactive waste management strategies worldwide.

In the history of British ILW policy one event stands out. On 17 March 1997, the day that the general election was announced, the Secretary of State for the Environment John Gummer upheld the findings of a local Cumbrian planning inquiry. The inquiry had found against the Nirex decision to build an underground geological laboratory known as the Rock Characterization Facility, or RCF, near Sellafield in Cumbria. The RCF was intended to take forward the Nirex programme of geophysics research from a programme of borehole drilling towards eventual repository construction and design. The Secretary of State's decision not only ended plans for the RCF but also stalled all progress on the management of ILW in the UK.

Since 1997, the Nirex programme of site investigations has been wound down. Long-term monitoring has continued, but boreholes have been progressively decommissioned [75]. Also since the RCF decision Nirex has shed approximately half its staff, as there has been relatively little work for the organization to do.

Only after a delay of several years did British government policy for ILW start to move forward again with the publication in late 2001 by the Department for Environment, Food and Rural Affairs (DEFRA) of a consultation document entitled *Managing Radioactive Wastes Safely* [76]. The emphasis on the consultation was to garner views on how best to develop a process through which nuclear waste policy could be established rather than to determine actual radioactive waste management strategies. A key recommendation from the restarted policy process was to create a new Committee on Radioactive Waste Management (CoRWM) to oversee the necessary processes of consultation for a management strategy and to determine a preferred option by 2006.

Despite the stalling of the Nirex programme in the spring of 1997, not all of its work to that point has been wasted. By the time of the RCF decision Nirex had spent more than £450 million on research including [77]:

- repository post-closure safety assessment,
- site selection and characterization,
- waste characterization and inventory compilation,
- provision of information to the public,
- computer modelling for safety assessment and commercial analysis,
- waste packaging design,
- waste transport design and
- repository design.

Depending somewhat on the strategy to be recommended by the CoRWM, much of the work done by Nirex will probably turn out to have been useful to those finally responsible for implementing Britain's preferred management strategy.

One key lesson to be learned by those planning a nuclear renaissance comes from a problem encountered by Nirex during the RCF planning inquiry. When Nirex needed independent experts to validate their scientific conclusions they found that there were very few UK geologists knowledgeable about radioactive waste issues who could be said to be truly independent of the organization. So many in the UK expert community had worked for, or received grants from, Nirex that there were almost no independent experts able to provide peer review. If there is to be a nuclear renaissance then there is a need for academic science and engineering research funders (such as the Research Councils) to support use-directed research in the area of nuclear fission and waste management. These individuals would be well placed to provide independent peer review when necessary.

While there is no evidence that the Nirex scientific programme has done anything other than serve the public interest, Nirex has frequently suffered from perceptions of conflicted interests. Because Nirex is owned almost entirely by the British nuclear industry (British Energy, BNFL and UKAEA) it is regarded by many as a creature of the industry. As a consequence its scientists have not been regarded as independent. It probably would have been preferable for Nirex to have been established as a wholly public body governed by the (Haldane) principle of best science. As it was, Nirex was constituted under the 'polluter pays' principle. That can be problematic when dealing with radioactive waste legacies as it is more difficult to persuade the public of the independence of an organization when it is funded by the polluters.

Another problem encountered by Nirex, and from which it has learned a great deal has been the issue of excessive secrecy. When Nirex first planned for an underground laboratory it had a list of 11 sites chosen for optimal geology and other technical characteristics. At a late stage the Sellafield site in Cumbria was added to the list and it emerged as the proposed site for the RCF. The locations of the other 11 sites have always been kept secret. Nirex maintain this secrecy arguing that the old list is now entirely irrelevant and that it would play no role in choosing future sites for any repository. To this one might reply, if the list has no relevance, why continue to keep it secret? However, publication might still have the power to blight property values near the sites on the list, and as a consequence publishing the list would likely be harmful. If there is to be major progress in UK waste policy it seems likely that it will not be on the basis of secrecy but on the basis of openness and transparency from the start. Nirex has adopted a transparency policy in response to its previous difficulties [78].

Contemporary with the reinvigorated British policy for operational radioactive wastes (the establishing of the CoRWM etc.) a start has also been made in connection with planning for the decommissioning of large-scale nuclear facilities, many of which will soon reach the end of their operational lives. Recognizing that it was the British people, as represented by Parliament, that made the key decisions concerning the development of the civil nuclear power programme (Magnox, reprocessing at Sellafield etc.), the government has acknowledged that the public purse should pay for the associated decommissioning. The British government has therefore proposed that a new public body, the Nuclear Decommissioning Authority (NDA), should be established to tackle these matters. As originally proposed, the NDA (originally known as the Liabilities Management Authority) was not to be responsible for the decommissioning of the British Energy AGR and LWR power stations as these liabilities were to be covered by a special segregated fund established as part of the privatization of British Energy. However, given the parlous financial situation of British Energy it seems possible that the NDA's remit might be extended beyond that originally envisaged and include responsibilities associated with the long-term liabilities of the company.

While the creation of a public body to manage public decommissioning liabilities might be regarded as a necessary step before the start of an industry-led nuclear renaissance, the name Nuclear Decommissioning Authority gives no hint that British policy-makers have any interest in new nuclear build. Rather the impression is to the contrary, hinting that the most likely scenario in the UK will be nuclear fade-out rather than any renaissance.

Responsibility for civil HLWs in the UK rests with BNFL. Nearly all these wastes arise from operation of the fuel-cycle. Little or no HLW is expected from decommissioning. As described earlier in this section, in the UK HLW is synonymous with the highly radioactive glass that emerges from reprocessing. The UK has no policy for the long-term management of these wastes. At present the vitrified waste is in secure storage at the BNFL Sellafield site. According to the 1998 Nirex DETR inventory, at that point 13% of Britain's 1800 cubic metres of HLW had been vitrified. Most observers expect that BNFL will complete vitrification of the vast majority of Britain's HLW in the years ahead. Once vitrified the waste is widely expected to stay in store for the decades required, following substantial radioactive decay, for it no longer to be self-heating.

In 1999 the then UK Department for the Environment Transport and the Regions identified a series of research questions would need to be addressed if HLW policy were to move forward in the UK [75]. These topics included:

1. Communication and risk perception
2. Waste forms, packaging, handling and transport

3. Repository design
4. Nuclear safeguards
5. Criticality
6. Heat emission
7. Vitrified waste behaviour
8. Spent fuel behaviour
9. Container evolution
10. Gas generation
11. Engineered barriers
12. Site selection
13. Site characterization
14. Safety assessment
15. Data, model and code availability
16. Biosphere and radiological risks and end points
17. Climate change
18. Underground laboratories
19. Repository geotechnical construction
20. Post-closure monitoring.

Furthermore, the 1999 DETR Research Strategy identifies 60 outstanding research topics required to address the 20 questions listed above [75]. Of these 60 research topics:

49% would be tackled by the policy implementer,
13% by the waste producers,
15% by policy makers and
23% by the regulators.

Importantly, 44 of the 60 topics would be suitable for international collaboration.

All of the research questions and topics identified by the DETR in 1999 are associated with the orthodox strategy of deep geological disposal, although some consideration is made of future waste definitions. No emphasis is given to nuclear waste transmutation or other radical approaches.

Even a relatively modest alteration of policy towards, for instance, underground storage with long-term retrievability (as recommended by a consensus conference organized by UK-CEED in 1999 [67]) would add to the range of HLW research needing to be addressed.

In summary, the UK government identified in 1999 that orthodox approaches to UK HLW management of legacy wastes would require substantial new research [75]. It seems likely that future thinking will extend far beyond the narrow remit highlighted by the DETR survey and as such the need for research and development will also be extended far wider. HLW is an area of UK waste policy where technical questions remain to be solved.

I.4.2.8 Other problematic radioactive materials

While not officially classified as wastes in the UK, these materials never-theless require careful long-term management. In those countries operating a once-through nuclear fuel cycle (e.g. the USA for its civilian fuel cycle, Switzerland and Sweden) spent fuel is the standard nuclear waste requiring management and is accounted for as such.

Spent nuclear fuel

The proliferation resistance arising from the hazards of manipulating spent nuclear fuel have led to the US suggestion that fissile materials should wherever possible be safe to the spent fuel standard. The fundamental assumption underlying this philosophy is that the fissile isotopes uranium-235 and plutonium-239 would be extremely difficult to remove from the fuel pellets inside the fuel rod because of the intense radioactivity coming primarily from the fission products in the spent fuel. Over hundreds of years this radioactivity decreases significantly and the proportion of pluto-nium that is fissile increases. This means that the spent fuel standard is not a good mechanism for proliferation resistance over the very long term, but in the short to medium term it is a simple and cheap way to ensure prolifera-tion resistance at the back end of the nuclear fuel cycle.

The spent fuel standard is often mentioned in the context of management options for separated plutonium. It is suggested that to improve proliferation resistance surplus separated plutonium should be mixed with high-level wastes to at least the spent fuel standard. For countries such as the UK with large legacy stocks of separated civil plutonium there is a high degree of irony in these ideas, because to dilute plutonium in HLW to the spent fuel standard would constitute the exact reversal of several expensive steps in the UK PUREX reprocessing strategy. For the UK there is insufficient untreated HLW with which to dilute separated plutonium back to the spent fuel standard, Strategies for the management of separated plutonium are discussed further later in this chapter.

Depleted uranium

The UK Department of the Environment Transport and the Regions estimated in 1999 that the UK currently has approximately 100 000 tonnes of depleted uranium [75]. While industrial uses for this material are limited, it is not regarded officially as a radioactive waste in the UK.

Naturally occurring radioactive materials (NORMs)

Although only rarely discussed, NORMs raise issues that can be challenging to many people's perceptions of nuclear issues. The key to this lies in the natural origins of these materials. In essence the problem becomes evident if one considers the proposal that if one were to dig up some naturally occurring radioactive material it could in principle be illegal to put it back exactly where you found it. Those motivated by considerations of natural law might regard this as an absurd situation, while those with a scientific mindset would feel that such regulations are both necessary and consistent with good policy. As the scientific mindset is the cultural norm within the nuclear industry, whereas the sense of a natural order is more prevalent in environmental campaign groups, one can end up with a situation where the nuclear industry is notably more cautious and concerned than those from environmental pressure groups.

The issues of NORMs are, however, more than simply philosophical. The oil and gas industries in particular extract significant amounts of NORMs in the form of sludges (oily and non-oily) and scales on the surfaces of pipe-work, valves and vessels [76]. Some of these NORMs are sufficiently radioactive that their management is regulated and controlled. Other lower level wastes are specifically exempt from the need for management and classification by the terms of the Radioactive Substances Act 1993. It is predicted that, as the oil and gas industries develop, the quantities of NORMs expected to arise will exceed the capacity of the current management arrangements. Significant effort will probably be needed if these wastes are to be managed with the level of care and attention as that applied to similarly hazardous radioactive wastes from the nuclear power industry.

It is clear that several problematic materials (spent fuel, depleted uranium, NORMs and plutonium) are not yet classified as waste in the UK. If it is necessary that radioactive materials be classified rigidly as wastes or assets then there is a strong case for reclassifying all these problematic materials as wastes. This author, however, is of the view that the accountant's requirement for a firm distinction between waste and asset is best not applied to any materials associated with the back end of the nuclear fuel cycle. In part this is because the determination of true status is in several cases extremely difficult (although I have declared my view that waste is a better label than asset); but in part also because in terms of management strategies the distinction is unnecessary once a philosophy of monitored retrievability is adopted in place of disposal (see text box: *A British paradigm shift?*).

I.4.2.9   Issues of decommissioning

To date the bulk of British concern for radioactive waste management has been devoted to the operational wastes arising from the nuclear fuel cycle. Increasingly, however, attention is focusing on the decommissioning of plant and equipment associated with the first and second generation of nuclear power plants. Figure I.4.6 illustrates the proportionate growth of the contribution of decommissioning to low level wastes.

A British paradigm shift?

In the years since 1997, the paradigm of deep underground disposal is gradually being replaced by one of monitored and retrievable underground storage. Nirex has effectively adopted this change in outlook, albeit with the proviso that there should be plans in place to seal the repository for 'disposal' after a few hundred years. Following the events of 11 September 2001, there is also more widespread acceptance that the storage of nuclear wastes should be underground and, quite probably, deep underground.

The pressures in favour of monitored retrievability are more sociopolitical than technical. The public (as assessed by consensus panels) clearly prefer that wastes should be accessible, monitored and retrievable, even if backfilling the repository with concrete (the disposal concept) is actually believed by experts to be safer.

If a monitored and retrievable deep underground store were adopted, then the nuclear waste debate would likely move forward in three important ways.

1. **The balance of expert and public scrutiny would shift away from the long-term issues of geology and hydrology towards greater consideration of the design and construction of waste containers.** Associated with this would be a move towards materials science and metallurgy. It seems that in shifting the balance of scientific inquiry to such matters, higher degrees of certainty and consensus should be achievable.

2. **Britain would avoid the need to answer the difficult question:** *what is waste?* Currently, and controversially, the UK classifies neither spent reactor fuel nor separated civil plutonium as waste. This means that these problematic materials have been excluded from the mainstream of nuclear waste debate in the UK. In 1999 the House of Lords Science and Technology Select Committee called for surplus separated civil plutonium to be declared waste [2]. According to a February 1998 Royal Society report [3], it is estimated that by 2010 the UK will have over 100 tonnes of separated plutonium representing approximately two-thirds of the world's total. As the UK has only one civil power station straightforwardly capable of utilizing uranium–plutonium mixed oxide fuel, much of the separated plutonium would seem to be surplus to domestic requirements. Monitored and retrievable deep underground storage would seem to be the best policy for plutonium until such time as it may be needed.

Similarly spent fuel from Britain's pressurized light water reactors is not officially waste and no long-term management strategy has been adopted. This includes the spent fuel cores from Britain's 27 nuclear submarine reactors, both operational and retired. Whether reclassified as waste or not, such spent fuel would benefit from long-term deep underground storage.

3. **There would be no requirement to complete the science.** In the traditional paradigm of deep underground disposal, with the repository sealed forever, both the experts and the public need to be confident that the scientific knowledge underpinning the policy is 100% robust and complete. While many experts take the view that such certainty is not far away, many members of the public doubt that such certainty will ever be achievable. A policy of monitored and retrievable underground storage could be accompanied by a sincere and ongoing push for better science in the area of repositories and also a push into more radical alternatives. An ongoing research programme deliberately constructed to challenge orthodox thinking would bolster public confidence and act as a competitive motivation to those working on the deep store programme. Government could declare sincerely that if better ideas emerge national policy priorities could be altered at any time in response to such innovation.

Figure I.4.7 illustrates the related growth in the contribution of decommissioning to intermediate level wastes over the same, roughly 200 year, period.

By contrast, there is no expectation of any high-level (self heating) wastes arising from a decommissioning of the existing UK nuclear power infrastructure.

Despite taking a leading role in the privatization and deregulation of utility industries, Britain proposed in a Department of Trade and Industry White Paper of July 2002 to adopt a Government owned and controlled approach to the decommissioning of the nuclear legacy through the establishment of the Nuclear Decommissioning Authority (NDA) [79]. The enabling legislation for the NDA forms part of the Energy Act 2004.

The amounts of money involved are substantial (£47.9 billion undiscounted as at March 2002) although so are the timescales envisaged for the NDA's operations (approximately 150 years) [79]. As noted earlier, the public responsibility for the NDA is entirely proper as it was the representatives of the public in parliament who made most of the key decisions to develop the nuclear infrastructure that is now to be dismantled. Any nuclear renaissance will surely differ markedly from those historical decisions, both

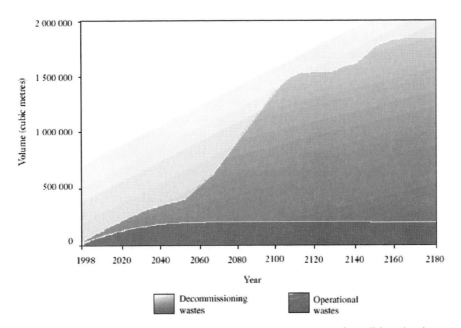

Figure I.4.6. Cumulative production of low level waste, in terms of conditioned volume [76].

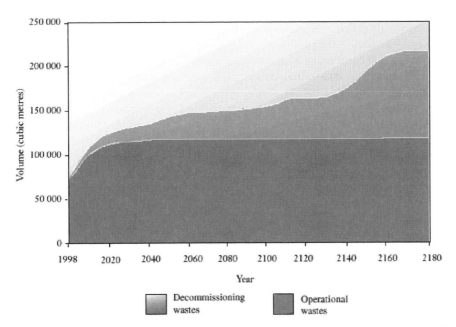

Figure I.4.7. Cumulative production of intermediate level waste, in terms of conditioned volume [76].

for the more business-like approach that would be adopted and for the relative absence of concerns for national industrial policy and national strategic defence factored into the earlier parliamentary decision making. The establishment of the NDA is an important step. Despite the fact that its name implies a phased end to nuclear power in the UK, it can also be regarded as a necessary step to allow for a nuclear renaissance in the UK founded on business-like and transparent decision making. The nuclear legacy will inevitably be very different from our nuclear future.

I.4.2.10   The UK plutonium problem

Within the UK the most problematic of the unofficial wastes is surplus separated civil plutonium. Britain is unique in its civil plutonium problem for two reasons, both of which have their roots in British 1940s attempts to produce a plutonium-based atomic bomb. The first decision was the decision to develop a relatively low temperature (albeit significantly hotter than the Windscale Piles discussed earlier) graphite-moderated reactor design with relatively low fuel burn-up (Magnox) the second was to adopt a fuel cycle based upon aqueous reprocessing and plutonium separation. The graphite moderators and the reprocessing were attractive options to early UK policy makers given the vast sunk efforts in terms of finance and institutional learning invested in these technologies for weapons purposes. The conscious decision of the UK to produce large amounts of separated civil plutonium in its Magnox and AGR fuel cycles was accepted at the time because in the classical nuclear fuel cycle plutonium was intended to be the fuel of the future, powering a second wave of reactor technology—the fast reactors. However, several countries have abandoned such research in recent years, the UK having done so in 1994. It seems increasingly unlikely therefore that the UK nuclear fuel cycle will ever return to the form intended for it when the UK originally embarked on a programme of industrial scale reprocessing.

Primarily for economic reasons, but also as result of concerns for proliferation and waste management, it seems likely that any first wave of a nuclear renaissance would be based not upon reprocessing, but rather on the less complex once-through fuel cycle with spent fuel being regarded as a waste stream. This view is endorsed by the 2003 Massachusetts Institute of Technology report entitled *The Future of Nuclear Power*. This recommends that nuclear power be given active consideration for future electricity production, but only on the basis of a once-through fuel cycle [29]. However, as is discussed in chapters 7 and 8, one must note that technologies similar to reprocessing continue to figure in the longer-term plans of many nuclear policy makers.

Historians will probably identify the key point in the ending of the UK insistence that plutonium is an asset rather than a waste as being a 1999 House of Lords Science and Technology Select Committee report that called for surplus separated civil plutonium to be declared waste [80]. According to a February 1998 Royal Society report [81], it is estimated that by 2010 the UK will have over 100 tonnes of separated plutonium representing approximately two-thirds of the world total.

Unusually plutonium combines high half-life with high radioactivity (each plutonium decay triggers a cascade of relatively short half-life daughter decays). It is also a chemically toxic heavy metal with an ability to form a very wide range of chemical compounds. Plutonium, therefore, is often regarded as an extremely toxic substance. In fact there are far more toxic and troublesome isotopes associated with nuclear wastes, such as the long-lived fission products. For instance, plutonium and its dioxide are insoluble in water and so plutonium has little mobility in undisturbed geological disposal scenarios. All agree that the greatest threat to health from plutonium would arise from the inhalation of small dust particles of the metal or its compounds rather than from ingestion in food or water. The risks in long-term plutonium management are therefore primarily associated with accidental human intrusion or malicious actions rather than by natural dispersal over very long periods of time.

Concerns have been expressed in the UK concerning plutonium-contaminated ILW which is currently not distinguished from other types of ILW. In particular concern has been expressed about the behaviour of plutonium-contaminated ILW in water which might act as a moderator for neutrons emitted the plutonium favouring fission and even possibly criticality incidents. Derek Ockenden, formerly on BNFL, specifically warned the UK-CEED consensus conference about the problems of plutonium-contaminated wastes and water intrusion [84].

I.4.2.11   Why is separated civil plutonium a problem?

The answer to the question above can be stated in one word: proliferation. If one accepts that separated civil plutonium is a proliferation hazard, as the United States has done since the mid 1970s, then policy for its future management becomes severely constrained. Some in the UK nuclear community, however, persist with the view that the main barrier to weapons proliferation (particularly of plutonium-based weapons) lies not in controlling access to the necessary nuclear materials but rather in controlling access to weapons designs and technical know-how. Few in the US would be so sanguine and there access to civil plutonium is itself regarded as a central proliferation risk (see text box *Civil and military plutonium and uranium*). Civil plutonium-based weapons would be physically large and have relatively low explosive yields, but experts do believe that they could be made to work. This prospect and the public panic that would surely be associated with any serious terrorist attempt to develop nuclear weapons (including those based upon civil plutonium) means that despite the technical difficulties

Civil and military plutonium and uranium

The key fissile materials for nuclear power and nuclear weapons are plutonium-239 and uranium-235. Their presence in nuclear fuels varies dramatically depending on particular circumstances. In the case of uranium, the isotopic mix is a function of the enrichment processes used. Enrichment applies techniques such as gas centrifuge technology or gaseous diffusion technology. In each case the fundamental principle is to separate lighter molecules of uranium hexafluoride (containing uranium-235) from the heavier molecules having only uranium-238 atoms. Further details are available from the US Nuclear Regulatory Commission's website (as of June 2004: http://www.nrc.gov/materials/fuel-cycle-fac/ur-enrichment.html#2).

Plutonium is obtained via the reprocessing of spent nuclear fuel. The isotopic composition of plutonium depends upon the technical details of the reactor, its operating temperature (normally fixed for a given design) and the length of time that the fuel was in the reactor core. These factors may be summarized by specifying the reactor design and the degree of nuclear fuel burn-up measured in megawatt days per tonne thermal power. Plutonium is made from nuclear processes in the reactor core affecting the fertile uranium-238 atoms in the nuclear fuel. While a distinction is often made between weapons grade plutonium and civil plutonium it is important to note that there is significant variation between different civil plutoniums depending on the design and operation of the source reactor.

While not without controversy it is widely accepted that civil plutonium (in both metal or dioxide form) could be made into a nuclear weapon. Such a weapon would be large and have a relatively small (and probably unreliable) explosive yield, but it could work [82]. Table I.4.2 illustrates how the proliferation risk from civil plutonium is particularly acute for British Magnox plutonium with its high fraction of fissile plutonium-239. Furthermore the nature of the Magnox design and the fact that reprocessing was required because of the fuel design leads Britain to have very large amounts of separated civil plutonium, estimated by the Royal Society to reach more than 100 tonnes by 2010, representing two-thirds of the world's total [81].

The attributes of uranium for civil and military use are classified according to the degree of enrichment used. As with plutonium, the fraction of fissile isotopes needed for a workable weapon is lower than that actually used in the construction of military nuclear weapons. Generally the term highly enriched uranium (HEU) is reserved for such weapons-usable material. The label HEU applies to all uranium where

Table I.4.2. Isotopic compositions of military and various civil plutoniums with associated burn-up assumptions [81].

Source of Pu (typical burn-up)	^{238}Pu	^{239}Pu	^{240}Pu	^{241}Pu	^{242}Pu
Weapons (400 MWd/t)	0	93.0	6.5	0.5	0
Magnox reactors (5000 MWd/t)	0	68.5	25.0	5.3	1.2
PWR reactors (33 000 MWd/t)	2.0	52.5	24.1	14.7	6.2
AGR reactors (18 000 MWd/t)	0.6	53.7	30.8	9.9	5.0

the proportion of uranium-235 has been enriched from its natural level of 0.7% to more than 20%. Actual nuclear weapons generally use HEU with enrichment levels around 90%. The term low enriched uranium (LEU) refers to uranium enriched to levels below 20% uranium-235. Typically LEU is between 2 and 6% uranium-235. It is important to note that LEU is not weapons-usable without further enrichment [83], a highly technically sophisticated process.

It is also important to stress that the civil nuclear fuel cycle can operate without any production of HEU and separated plutonium. Furthermore, HEU has had a negligible role in the history of civil nuclear power.

involved, separated civil plutonium must be regarded as an attractive material for technologically capable terrorists or rogue states.

Given that the UK has a large stock of separated civil plutonium, and others (notably the Russian Federation) have large stocks of military grade plutonium, what long-term management options are available?

I.4.2.12   Plutonium utilization in fast reactors

Despite most countries, except Russia and Japan, turning away from fast reactors in recent years, the possibility remains that surplus plutonium could be used to fuel a new generation of fast reactors (see chapters 7 and 8). While possible, it seems on balance unlikely that this would form a significant part in any early phase of a nuclear renaissance. In many parts of the world the momentum of the fast reactor programmes has been lost. It will probably take large-scale international initiatives such as the US-led Generation IV project to slowly restore such momentum. Certainly in Europe and North America fast reactors would seem to be off the agenda for the time being.

I.4.2.13 Plutonium utilization in mixed oxide (MOX) thermal reactor fuel

At the end of the 20th century this was the dominant proposal for the utilization and elimination of Britain's plutonium stocks. The plan was to mix it in with uranium to form a new type of reactor fuel known as mixed-oxide or MOX nuclear fuel. At present, however, no British reactor is licensed to operate with such fuels and only one (Sizewell B) could do so straightforwardly. It may be possible to operate the UK fleet of Advanced Gas-cooled Reactors on MOX fuel but the necessary safety case work would be substantial and has not been attempted anywhere.

While a MOX plant has been licensed for operation at Sellafield, its current role is for the production of fuel using overseas plutonium temporarily held in the UK as a result of international reprocessing contracts. It is far from clear that the Sellafield MOX plant activities will ever extend to include domestic British plutonium as there is no domestic route for the use for such fuel. Incorporating foreign plutonium in MOX fuel for return to its country of origin avoids the need to ship separated plutonium dioxide or plutonium metal internationally. While, as we shall discuss, there is little technical difficulty in removing plutonium from a freshly fabricated MOX fuel rod, if plutonium is shipped in that form, at least it is not usable for terrorist purposes without significant preparatory work. There is a danger, however, that the need for such work might be interpreted too strongly as being a significant source of proliferation resistance.

Some advocates for MOX shipments have declared that such shipments are intrinsically resistant to diversion by terrorists seeking to acquire weapons-usable plutonium. They assert that newly fabricated MOX fuel, such as might be shipped internationally, is inherently resistant to separation into its chemical components (such as plutonium) because of the overwhelming radioactivity of the other components of the fuel. Such views are, however, unduly sanguine. While a spent fuel rod *arriving* at Sellafield *for* reprocessing may have an intrinsic resistance to terrorist diversion because of the high radioactivity of the fission products and the minor actinides created during the use of the fuel rod, a new MOX fuel rod *leaving* Sellafield *after* reprocessing and fabrication would have only the very limited radioactivity arising principally from the plutonium component (particularly plutonium-241). If the terrorists were capable of assembling a nuclear weapon, then the modest hazards of working with civil plutonium would seem not to present an obstacle to them, especially if they were to care little for their own health. The chemical processes required to separate plutonium from a new MOX fuel rod are significantly less challenging than the equivalent task of separating plutonium from very highly radioactive spent fuel. The shielding required even by suicidal terrorists in any attempt to manipulate spent nuclear fuel would be so substantial as to hopefully render their plans completely impractical. New MOX fuel is very different and it would, therefore, seem correct that those who criticize a market in MOX fuel as adding to the risk of nuclear weapons proliferation deserve some consideration.

The separation of plutonium in reprocessing so that it may be used in MOX fuels has been described as 'spherically daft' [85], in that the process seems unwise from whatever perspective it is viewed, be it economics, environmental, safety, proliferation resistance, energy security, or any other.

Another key problem for MOX has been the low market price of new uranium for nuclear fuel (see figure I.4.5). It was originally believed in the early 1970s that large-scale nuclear fuel reprocessing could be economic if implemented in large-scale facilities such as the Thermal Oxide Reprocessing Facility (THORP) at Sellafield. The uranium extracted during reprocessing (which itself requires re-enrichment, as the unused uranium-235 in spent fuel uranium is insufficient to power a light water reactor without being topped up) was to compete on cost with new uranium. Since the 1970s, however, the anticipated expansion of nuclear power has not occurred and demand for uranium has not risen. Simultaneously supplies of new uranium into the market have been larger than had been expected, notably from non-mined supplies arising in part from the dismantling of the Soviet Union's nuclear arsenal. This has resulted in low uranium market prices and hence significantly poorer economics for reprocessing than had been anticipated when the key decisions to develop large-scale reprocessing were made. If conventional uranium-based fuels produced via reprocessing and re-enrichment are uneconomic, then MOX fuels are even more so, given the extra complexity involved in the plutonium processing, in the fuel fabrication and in reactor operation, as safety tolerances are tighter for MOX utilization. Any holistic assessment of the industry must concede that currently MOX fuels are significantly more expensive to produce than original nuclear fuels manufactured from new uranium.

As the UK has only one civil power station straightforwardly capable of utilizing uranium–plutonium MOX fuel, much of the large stock of UK separated civil plutonium would seem to be surplus to domestic requirements. As the isotopic mix of civil plutonium alters with time it is most straightforward and cost effective to have the most rapid MOX fuel fabrication possible. One issue is that plutonium-241 decays with a half-life of 14 years to form longer-lived americium-241. Americium isotopes are to be avoided in thermal reactor fuels (such as conventional MOX fuels) as they are strongly neutron absorbing and yield intensely radioactive curium isotopes when exposed to the thermal neutron flux. The chemistry and engineering is such that MOX fuel manufacturers can work around these difficulties, but it is clearly more straightforward for them if little americium is present in the feedstock coming into the MOX manufacturing process. These issues are discussed further in chapter 7 in the context of the transmutation of nuclear wastes.

If MOX were to be implemented as a solution to the British separated plutonium problem in the near future then one of the following options would seem to be needed. First, UK MOX might be exported as part of an international trade in plutonium-based nuclear fuels. Given the increased risk of diversion by terrorists and the likelihood that international demand will be limited, this possibility is looking increasingly unattractive and unlikely. Second, the MOX fuel might be used in a new fleet of British pressurized water reactors, such as the BNFL Westinghouse AP1000 design (see chapter 5). Third, with some work it would seem to be possible to utilize MOX fuel in the existing fleet of AGRs. This would require significant research and regulatory oversight as AGRs fuelled with MOX would have different reactor control properties compared with conventionally fuelled operations. The concerns in that case would seem to be cost and safety. Another issue is that the AGR fleet is owned and operated by British Energy. British Energy was established as a private utility free of government control. In recent years its financial position has become so perilous that it has become beholden to government for support. Such renewed ties to government might make it easier for an AGR MOX programme to be adopted in the UK. At the time of writing, however, no plans seem to be being made for UK MOX utilization in the AGR fleet. Given the limited remaining life of the UK AGR fleet and the significant regulatory and technical obstacles that would face any move towards AGR MOX fuel it seems extremely unlikely that such a course of action would be judged to be worthwhile.

Given the difficulties facing all possible short-term uses for UK MOX, monitored and retrievable deep underground storage would seem to be the best policy for separated plutonium until such time as it may be needed. One example of such a long-term need for MOX would be if a large fleet of new water-cooled power reactors were to be constructed as part of a nuclear renaissance in the UK.

Some argue that eventually there will inevitably be a return to a desire to utilize currently surplus plutonium and that this possibility should motivate monitored retrievability for any long-term plutonium management strategy.

I.4.2.14 Deep geological disposal

The long timescales associated with the radioactive decay of most plutonium isotopes suggests the possibility of disposing of unwanted plutonium deep underground in a facility similar to that proposed for more conventional high-level wastes. While currently the volume of plutonium in monitored storage is rather small (plutonium is very dense) the level of dilution necessary for safe geological disposal would make the task of geological disposal a substantial one. In order to reduce the long-term proliferation risks discussed earlier, plutonium for deep underground disposal would probably be mixed with conventional HLW, although for the UK this could be difficult given the lack of raw HLW with which to dilute separated plutonium. To date there has been little research in the UK into the geological disposal of plutonium. If surplus plutonium were to be regarded as a high-level waste to be disposed of with other types of HLW, one problem would need to be overcome. At present several countries, including the UK, vitrify HLW into borosilicate glass blocks. Unfortunately for chemical reasons the glasses used are unable to accommodate significant quantities of plutonium [86]. If plutonium were to be managed as a waste for disposal, new waste forms with long-term stability would need to be developed.

Earlier in this chapter it was proposed that it is probably unhelpful to force the classification of problematic radioactive materials into one of two categories, namely, waste or asset. This is particularly true for surplus separated plutonium. Even if policy makers are forceful and robust in deciding, for instance, that such plutonium is a waste and that it should be disposed of deep underground, or that it is an asset and should be stored for future use, the risk of long-term problems would remain.

It is an unfortunate likelihood that over the long timescales of nuclear waste management (plutonium-239 has a half-life of 24 000 years) some currently stable industrial societies will descend into anarchy and civil strife. Already there have been major wars in areas with nuclear infrastructure such as the Balkans and Iraq.

In the event that a failed semi-medieval state arises in an area where a policy of deep geological disposal of plutonium had once had been adopted, there is a risk that some future warlord might try to dig plutonium mines to retrieve the legendary material from where it had been buried hundreds of years before. It is certain that in order for fission weapons to be constructed, such a warlord would need access to reasonably sophisticated chemistry and materials facilities, but these are not far beyond the resources available to failed states today. The alternative of indefinite long-term storage of plutonium might, as society slowly crumbles, give rise to, in Thomas A Sebeok's memorable phrase, an 'atomic priesthood'. The priesthood would retain the knowledge and skills within their tight community of scholars while the society around them cares little for knowledge or science. The metaphor is powerful as one looks back to the monks of medieval Europe carefully husbanding literacy and wisdom through the Dark Ages.

I.4.2.15   Incineration as part of a waste transmutation programme

It seems that there are serious and special difficulties with both the storage and the disposal of plutonium. While these difficulties are surely not insurmountable they are such that one is drawn to consider more radical and innovative approaches. Among such ideas one seems to have particular merit: the use of particle accelerator technology to transmute problematic plutonium into more easily handled materials. This modern form of alchemy, that would have the added benefit of generating electricity, is discussed in depth in chapter 7.

I.4.3   The US experience

Noteworthy and somewhat unexpected progress on nuclear waste management has been made during the presidency of George W Bush. As with so much of his policy agenda the unanticipated trigger for action has been the terrorist attacks of 11 September 2001, and the subsequent War on Terror.

In the United States the Department of Energy (DOE) is responsible for the management and safe disposal of spent nuclear fuel from commercial nuclear power plants. In 1983 the Nuclear Waste Policy Act was signed into law by President Reagan. It established the Office of Civilian Radioactive Waste Management within the DOE and it authorized the development of a repository for high level wastes including spent fuel. In 1987 Yucca Mountain was designated as the site for the repository by the Nuclear Waste Policy Amendments Act. Shortly after the 11 September attacks, and with greatly heightened concern in the United States for the possibility of radiological or nuclear terrorism, President Bush formally recommended the Yucca Mountain site to Congress on 15 February 2002. Governor Kenny Guinn immediately issued a formal notice of disapproval to Congress reflecting local objections from the state of Nevada where the Yucca Mountain facility is located. On 18 May 2002 the House of Representatives approved the Yucca Mountain Repository Site Approval Bill by a 306 to 117 majority. On 9 July 2002 the Senate indicated its approval for the Bill with a majority of 60 to 39. On 23 July 2002 President George W Bush signed the related Joint House Resolution 87.

Although one might criticize the democratic legitimacy of the George W Bush presidency given the extremely close election result in 2000 and the difficulties in Florida, it is clear that the recent US experience is one of top-down leadership from central government. While the Yucca Mountain process is far from over, and there is much associated activity in the US courts, it seems clear that Napoleonic approaches to such contentious issues can generate progress if the socio-political circumstances so permit. As we shall see this contrasts very strongly with similarly effective measures in Finland in recent years.

I.4.3.1   US waste classifications

The US maintains four categories of radioactive waste: high level waste (HLW), transuranic waste (TRU), low level waste (LLW) and mill tailings.

We shall consider these waste categories in turn as in each case they differ in definition and in physical waste type from their nearest UK equivalents.

High level waste is defined by the US Nuclear Regulatory Commission (NRC) as: irradiated (spent) reactor fuel, liquid wastes from reprocessing solvent extraction, and solids from which such liquid wastes have been converted [87]. The clear view of the US NRC is that spent nuclear fuel is a high level waste. This policy axiom runs entirely counter to the official view in the UK where spent fuel is not classified as a waste, but it is consistent with policy in other countries, such a Sweden and Switzerland, that have also adopted a once-through fuel cycle. Neither the NRC nor the US Department of Energy specifies the specific radioactivity required for wastes to be deemed HLW [87].

Within the US nuclear policy community the DOE has always been somewhat reluctant to follow the lead of US high-level policy-makers and regard spent fuel as a radioactive waste. Its own public-sector spent fuel inventory is often excluded from estimates of its radioactive waste legacy. Separately, the US DOE has successfully pushed for the creation of a new US category of radioactive waste known as transuranic waste (TRU).

Formally the DOE defines TRU wastes to be those containing more than 100 nanocuries of alpha-emitting isotopes per gram, with atomic number greater than 92 and with half-lives greater than 20 years, excluding high-level wastes and some specifically exempted wastes [87]. TRU is a product of the US reprocessing activities associated with nuclear weapons production. The DOE is disposing of military TRU waste at the Waste Isolation Pilot Plant (WIPP) near Carlsbad, New Mexico. The WIPP facility is noteworthy for having been constructed in a large natural salt formation. The presence of the salt indicates that the local geology has been dry for millennia and also salt has the benefit that it is self-healing to fractures. WIPP is expected to receive 6 million cubic feet of wastes during operations [88]. Because TRU lacks the self-heating decays from fission products it is more easily handled than HLW. It is also well suited to waste transmutation in either thermal neutron or fast neutron based nuclear waste burners (see chapter 7).

With the exception of mill tailings LLW is the category applied to all US wastes that are neither HLW nor TRU [87]. US LLW generally has shorter half-lives than either HLW or TRU.

Uranium milling consists of the mechanical and chemical processes by which raw uranium ore from the mine is processed into dry yellowcake ready for transport to enrichment facilities or, increasingly less commonly, directly to fuel fabricators. Chemically yellowcake is a uranium oxide (U_3O_8) powder that unsurprisingly is a rich yellow in colour.

Uranium mill tailings are substantial in volume as many uranium ores contain only about 1% uranium. Recently mining in Canada has started to extract uranium from ore seams that are as rich as 20% uranium, and this presents serious technical safety-related challenges.

As the Sweetwater Uranium Mill in Wyoming, USA, explains, the construction, operation and reclamation of facilities for uranium mill tailings are regulated by both the US Environmental Protection Agency (EPA) under 40 CFR Part 61 Subpart W (NESHAPs) and by the US Nuclear Regulatory Commission (NRC) under 10 CFR 40 (Domestic Licensing of Source Material). In addition, various NRC regulatory guides, staff technical positions (STPs) and other documents also regulate the construction, operation and reclamation of uranium mill tailings facilities (see: http://www.wma-minelife.com/uranium/mill/mllframe.htm).

I.4.4   The Finnish experience

In contrast to the United Kingdom, Finland has been able to make good progress on the radioactive waste question. In contrast to the United States this has been achieved via a process sensitive to, and driven by, local considerations.

Recent Finnish progress on nuclear matters goes beyond waste management to include the core issue to nuclear renaissance—new nuclear build. In May 2002 the Finnish parliament ratified an application from the nuclear plant operator Teollisuuden Voima Oy (TVO). TVO already operates two western designed boiling water reactors built by ABB Atom AB with a combined output capacity of 1680 MWe. These plants are located at Olkiluoto on the western coast in the south of the country. The successful proposal by TVO to construct a new plant (the fifth in Finland) represents the first nuclear plant to be proposed in either western Europe or North America in at least a decade. Importantly the Finnish policy processes associated with the development of new build and for consideration of the fuel cycle of the new plant have been kept completely separate from decisions concerning the management of wastes arising from the four existing plants.

The Finnish radioactive waste burden consists primarily of high level wastes in the form of spent fuel and low level wastes that will arise in the eventual decommissioning of the existing plants. At the end of 2001 the Finnish high-level waste legacy consisted of two spent fuel waste streams. The first consists of 5274 bundles of BWR fuel equivalent to 923 tonnes of fresh uranium (burn-up between 23 and 40 MWd/kgU) from the TVO Olkiluoto plants. The second HLW stream comes from the two Russian designed VVER-440 V213 pressurized water reactors operated by Fortum Power and Heat Oy at Loviisa on the coast in the south-eastern corner of the country. The VVER waste legacy consists of 2335 bundles of VVER-440 fuel equivalent to 279 tonnes of uranium with a burn-up of 34–38 MWd/kgU. The Finnish HLW legacy is increasing by about 256 bundles of BWR fuel and 204 bundles of VVER fuel each year [89].

The success of Finland in managing its radioactive waste legacy is most often associated with a Decision in Principle to construct a waste repository at Olkiluoto. That decision was ratified by the Finnish parliament in May 2001 with a vote of 159 in favour and three against. The path to the successful parliamentary vote is an interesting one, and one that contrasts greatly with the US experience.

In 1983 the national site selection process began for a high-level waste repository, the government setting a deadline of 2000 for final site selection. By 1992 an original list of 102 potential sites had been whittled down to just the three most appropriate locations to be investigated in detail. In 1997 Loviisa was added to the short-list following the ending of shipments of spent fuel to Russia for reprocessing. In May 1999 the radioactive waste research and development company Posiva Oy (jointly owned by TVO and Fortum) submitted a Decision-in-Principle (DiP) application to the government suggesting Olkiluoto as the preferred site. The following year the Radiation and Nuclear Safety Authority (STUK) submitted a preliminary appraisal of the proposal. Separately Eurajoki, the local municipality of the Olkiluoto site, gave its approval to the DiP application. Two local residents appealed against this decision, although these appeals were later rejected by the Supreme Administrative Court. In December 2000, consistent with its own original deadline from 1983, the government made a Decision in Principle in favour of the Olkiluoto repository plan and then referred the matter to parliament. The required parliamentary endorsement was received in May 2001. One year later parliament endorsed the separate new nuclear build proposal. In June 2002 a survey undertaken by Suomen Gallup for the Central Organization of Finnish Trades Unions (SAK) showed that 55% of the Finnish population supported the new nuclear build and, by implication, its first steps towards a nuclear renaissance.

It is remarkable that Finnish policy for these thorny nuclear issues managed to move forward on schedule and without severe disruptions. The key attribute of the Finnish process for the management of radioactive wastes is a policy of transparency, community volunteerism and local engagement. From the start, the project developers sought to explain their ideas to the local community. To illustrate the commitment to the community Posiva Oy undertook to move its corporate headquarters to the small town of Olkiluoto (population approximately 6000). The corporate HQ was to be a manor house on the edge of the town that up until that point had been used as a retirement home. Posiva Oy compensated the townspeople for the loss of their retirement facility by constructing for them a brand-new state of the art complex. Whether the compensation was also for the burden of being host to a radioactive waste repository is an ongoing subtext, but one that does not apply formally. That is because at any point the townspeople had the right enshrined in law to block further progress on the repository plan. The Finnish strategy was based on trust by the nuclear policy makers of the local community and on trust by the community for the safety of the new facilities proposed by the policy-makers.

The trust-based models successfully adopted in Finland would seem to be a model for progress in other countries.

While the Finnish approach based upon mutual trust is in complete contrast to the British experience of widespread mistrust and failed policies, there is one aspect of the Finnish and British approach that was similar and contrasted with the US experience. Both Olkiluoto in Finland and Sellafield in the UK (the proposed site for the failed Rock Characterization Facility proposal) have long-standing associations with the nuclear power industry. By contrast the state of Nevada (home to the proposed Yucca Mountain repository) has no history of civil nuclear power production. The Nevada links to nuclear issues are restricted to being the location for numerous nuclear weapons test during the Cold War.

In the UK the Sellafield option was added to the site list for the proposed RCF relatively late in the day on the basis that the site had a nuclear heritage. In the British experience this aspect has turned out to be insufficient to ensure policy success. In Finland, however, the long-standing familiarity by the Olkiluoto community with the civil nuclear power industry has undoubtedly helped with their enthusiasm for the repository project. It seems likely that communities with a positive nuclear heritage should be considered when seeking to advance the cause of nuclear renaissance within the area of waste management or in the case of new build projects. A process of genuine engagement between nuclear policy makers and concerned publics is essential if progress is to be made.

I.4.5   The need for dialogue on waste in any nuclear renaissance

By the late 1980s and early 1990s most technical commentators on nuclear waste management have tended to take the view that the technical issues had been resolved satisfactorily and that the only remaining challenges were social and political. Tang and Saling reflect the attitude of the nuclear waste professional well when they say [90]:

> ... the public is apprehensive about the potential hazard of radwaste material. The importance of understanding the management technologies, design philosophy and performance evaluation related to processing, packaging, storing transporting, and disposing of radwaste cannot be overemphasized and such information must be disseminated.

Insight from those working in the public understanding of science has revealed, however, the weakness of such 'deficit model' approaches. In this context the deficit model holds that if the public could only be educated as to the realities of nuclear power (and its wastes) then they would cease to be anxious about it and would begin to accept the strategy proposed by the technologists. In fact, studies have shown that such educational initiatives tend to result in a hardening of pre-existing attitudes. Those who are hostile to nuclear power are confirmed in their opposition by the information provided. Those who are positive about the technology are similarly confirmed in their own view by the same information.

At the start of the 21st century a new approach is emerging that seems far more likely to yield results in this hugely contentious area. The old paradigm of education and public understanding is being replaced by genuine dialogue and desire for consensus. It seems likely that the best hope for real progress lies in a genuine multi-way communication among the professionals tasked with nuclear waste management, the public (both locally and nationally), and environmental organizations. The loudest voice in this process should reside with the concerned publics if real progress is to be made. Adopting such an approach is not a threat to the nuclear technologists; in fact in many cases it represents an opportunity for progress. For instance, the UK's move away from prompt geological disposal of wastes towards monitored retrievability came originally not from technological analysis but from public insistence (see text box: *A British Paradigm Shift?*).

Also important is the need for technologists and policy makers to listen when involved in such dialogue. The troubled history of nuclear waste management in several countries includes a failure to listen and to look for opportunities in the public attitude. I suggest that:

> *When the scientists and engineers discussed nuclear power with the public, the members of the public said that they were scared of the dangers of nuclear power.*

For decades, scientists and engineers worked at great expense to reduce the danger from nuclear power, yet only in recent years has the industry started to pay attention to the more socio-politically important fact that people are scared. The most effective route to minimize the fear is not to minimize the danger, as there are other factors such as fairness, volunteerism, timescale of hazards and familiarity that are far more important to reducing fear. Any nuclear renaissance will need to take intelligent account of these aspects if it is to make progress. (See text box *Public Attitudes*.)

Chapter 7 deals with nuclear waste partitioning and transmutation (P&T) in some detail. Industry and public attitudes to P&T are, however, interesting in so far as they further illustrate key differences in regard to attitudes to danger or 'risk'. It seems likely that a P&T fuel cycle will expose a larger number of nuclear industry workers to possible danger over the decades in which such a facility operates. In contrast the deep geological disposal of spent fuel, ILW and HLW would inevitably expose very large numbers of members of the public, over the very long term, to infinitesimally

Public attitudes

It is clear that in many people the idea of nuclear waste arouses strongly negative feelings. These reactions have been found to be strongest for those closest to proposed nuclear waste disposal facilities [69]. The observation of vehement local opposition has given rise to the pejorative label NIMBY for 'Not in My Backyard'. From the 1940s until the 1980s all decisions about nuclear waste management seem to have been made by technical experts with physical science or engineering backgrounds. Their approach to technical questions was reductionist and axiomatic. They agreed that radioactive waste posed a hazard to health and the environment. The basis of the hazard was radiological and to some extent chemical toxicity. The key to a solution to the problem would be to reduce the total amount of danger in the system associated with the perceived hazard. The absence of a sensitivity to psychology and politics meant that the technical experts failed to pay sufficient attention to stress as a hazard and that they failed to appreciate that the public do not in fact regard all human lives as being of equal importance.

Politics and the importance of public attitudes in a democracy are every bit as real as the ionizing properties of high energy charged particles. If the technical experts, and more importantly their employing organizations, had been more receptive to such ideas far greater progress might have been possible by now.

From the late 1970s onwards much understanding has been gained about how the public responds to risks and hazards. These insights provide an opportunity for a nuclear renaissance in sympathy with public concerns. If the attitudes of ordinary people continue to be regarded as irrational or uneducated by those responsible for nuclear waste policy then little progress would seem to be possible.

The key factors shaping public attitudes to risk are as follows.

- Voluntarily assumed risks are preferred to imposed risks.
- Immediate consequences of accidents are preferred over equivalent delayed consequences.
- Familiar risks are preferred over unfamiliar risks.
- Accidents where individual actions can provide mitigation are preferred over uncontrollable risks.
- Old risks are preferred over new risks.
- A risk that harms a few people very often is preferred over one that harms very many people very rarely.

- Some risks have an attribute of being 'dread'; these can relate to complex social and cultural issues defining the nature of fear (religious, subconscious, artistic etc.).

Generally speaking nuclear power has scored extremely badly on all these criteria [70]. Such public attitudes favour approaches to radio-active waste management in which the balance of risk is borne by informed volunteers, such as professionals exposing themselves to the risk of industrial accidents. Furthermore it is not in the interests of those managing nuclear waste to seek to avoid discussion of the issues. Rather it is important to encourage the widest possible familiarity with the problem. The information needs to be both true and internally self-consistent. However, Malcolm Grimston and Peter Beck rightly highlight the difficulties of traditional approaches when confronting public concerns [19]. They report an awkward moment for nuclear waste policy makers when the debate threw up the question that if intermediate level radioactive waste is 'about as dangerous as paint-stripper' then why must it be buried one kilometre underground in solid granite?

small probabilities of harm. The multiplicative product of the probability of harm and the number exposed to that probability could actually be very similar in both cases. The nuclear industry is similar to all responsible large industries in seeking to minimize safety hazards to its work force. Therefore it tends to the view that it is probably worse to expose an identifiable individual to a measurably large probability of harm over the near to medium term than seek to eliminate the already infinitesimally small probability of harm to the wider population over geological timescales. The tiny risks from geological disposal fall upon those who will never be known to us; who will themselves never be able to tell what made them sick and which can never be associated with certainty to the cause of their difficulties. These few cases would be lost in the noise of the enormous number of people losing their lives for completely natural causes over this vast period. Those in the industry therefore usually tend to the view that the deaths of real people are the most important things to avoid. Conversely, however, from the public standpoint, harmful effects are often more acceptable if borne by informed volunteers over the near term (such as nuclear industry workers) rather than being imposed on innocent members of the public over the very long term [91]. Not all deaths are equal in the public consciousness and research suggests that, for the public, the deaths of real people is not necessarily the most important thing to minimize. Furthermore, the politics of the nuclear industry are special and there is a greater than usual need for the industry to be clear that its primary concern is public safety. Thinking about attitudes to P&T, and remembering attitudes to monitored retrievability of wastes versus geological disposal can lead one to the surprising conclusion the public can, on occasion, prefer a technology that the technical experts regard as unacceptably dangerous and verging on the professionally unethical. The idea that the public might call for more dangerous waste management strategy is a clear indication of the idea that the public do not seek a minimization of danger. What they seek is a minimization of fear.

Part I references

[1] McCarthy C, OfGEM Chief Executive, speech at the European Centre for International Research, *Regulating national energy markets within a European Framework: a regulator's perspective*, 26 March 2003

[2] MacKerron G 2004 *Energy Policy*, **32**, pp 1957–1965

[3] House of Commons, *Hansard*, 6 September 1990, Column 600

[4] van Oostvoorn F and Boots M G, *Impacts of Market Liberalization on the EU Gas Industry*, 1st Austrian-Czech-German Conference on Energy Market Liberalization in Central and Eastern Europe, Energy Research Centre of the Netherlands (ECN), ECN-RX-99-032, October 1999

[5] Fairley P 2003 'Recharging the power grid', *Technology Review*, March, pp 50–58

[6] Nuttall W J 2004 *The Engineer*, April 18, 19

[7] Patterson W 1999 *Transforming Electricity* (London: Earthscan)

[8] Hore-Lacy I 2003 *Nuclear Electricity* 7th edition (Melbourne, Australia: Uranium Information Centre)

[9] Department of Trade and Industry UK, Energy White Paper, *Our Energy Future— Creating a Low Carbon Economy* (London: Stationery Office), February 2003, Cm5761, page 11, paragraph 1.21

[10] Nuclear Energy Agency 2003 *Nuclear Energy Data 2003* (Paris: Organization for Economic Cooperation and Development)

[11] Department of Trade and Industry, UK, *Digest of United Kingdom Energy Statistics 2001* (London: The Stationery Office)

[12] Bilenky S M 2004 *Proceedings: Mathematical, Physical and Engineering Sciences*, The Royal Society, London, 8 February 2004, vol 2042, pp 403–443 (41)

[13] Kearns E, Kajita T and Totsuka Y 2003 'Detecting massive neutrinos', *Scientific American*, **289**(SPI1), pp 68–75

[14] Hewitt G F and Collier J G 2000 *Introduction to Nuclear Power* (London: Taylor and Francis)

[15] Nuclear Energy Agency 2000 *Nuclear Power in Competitive Electricity Markets*, Organization for Economic Cooperation and Development, Paris, p 10

[16] Department of Trade and Industry, Energy White Paper, op. cit. p 77, paragraph 6.6

[17] International Energy Agency 2001 *Nuclear Power in the OECD* (Paris: IEA)

[18] Economics Focus 2004 'Slackers or pace-setters?', *The Economist*, **371**(8376), 22 May, p 92

[19] Grimston M C and Beck P 2002 *Double or Quits—the global future of civil nuclear energy* (Royal Institute of International Affairs: Earthscan)

[20] Nuclear Energy Agency 2000 *Reduction of Capital Costs of Nuclear Power Plants* (Paris: OECD)

[21] The Lifestyle Movement, Bristol, UK (as of August 2004: http://www.lifestyle-movement.org.uk/str2/ethicalinvest.htm)

[22] Department of Trade and Industry, Energy White Paper, op. cit. p 107

[23] Bolter H 1996 *Inside Sellafield* (London: Quartet Books), p 54

[24] Imperial College—Centre for Energy Policy and Technology, *Assessment of Technological Options to Address Climate Change*, a report for the UK Prime Minister's Strategy Unit, December 2002

[25] Uranium Information Centre, *The Economics of Nuclear Power*, Briefing Paper 8, Melbourne, Australia, March 2004

[26] Report of the US National Energy Policy Development Group, Chair R Cheney, US Government Printing Office, Washington, DC, May 2001

[27] Energy Information Administration, US Department of Energy (http://www.eia.doe.gov/cneaf/nuclear/page/at_a_glance/reactors/wattsbar.html)

[28] McDonald A and Schrattenholzer L 2001 *Energy Policy*, **29**, 255–261

[29] Massachusetts Institute of Technology 2003 *The Future of Nuclear Power: an interdisciplinary MIT study*, Cambridge, MA, USA

[30] Environment Agency, *Trends in Air Quality* (www.environment-agency.gov.uk). Relates to: Environment Agency 2000 *The State of the Environment of England and Wales: the Atmosphere, 2000* (London: The Stationery Office)

[31] Nuclear Energy Institute, 'How Nuclear Energy Contributes to Clean Air Compliance', On-line column 'UpFront', Washington, DC (as of June 2004: http://www.nei.org/index.asp?catnum = 2&catid = 41)

[32] International Nuclear Safety Center, Argonne National Laboratory, Argonne, Illinois (as of June 2004: http://www.insc.anl.gov/pwrmaps/map/china.php)

[33] Energy Information Administration, US Department of Energy, *Timeline of the Chinese Nuclear Industry, 1970 to 2020* (as of June 2004: http//www.eia.doe.gov/cneaf/nuclear/page/nuc_reactors/china/timeline.html)

[34] de la Chesnaye F *et al* 2001 *Energy Policy*, **29**, 1325–1331

[35] Babiker M H *et al* 2002 *Environmental Science and Policy*, **5**, 195–206

[36] Davidovits J and Davidovits F, *Up to 80% Reduction of CO_2 Greenhouse Gas Emission During Cement Manufacture* (Saint-Quentin, France: Institut Géopolymère) (http://www.geopolymer.org/research_projects/global_warming_greenhouse_gas/cement_reduce_CO2_emission.html)

[37] Nuclear Energy Agency 2002 *Nuclear Energy and the Kyoto Protocol* (Paris: Organization for Economic Cooperation and Development)

[38] International Panel on Climate Change, downloadable graphics (www.ipcc.ch)

[39] Hoffman P 2002 *Tomorrow's Energy, Hydrogen, Fuel Cells and the Prospects for a Cleaner Planet* (Cambridge, MA: MIT Press)

[40] Schultz K, General Atomics, Presentation to the Stanford Global Climate and Energy Project, 14 April 2003

[41] Mathias P M and Brown L C 2003 *Thermodynamics of the Sulfur–Iodine Cycle for Thermochemical Hydrogen Production*, Presentation at the 68th Annual Meeting of the Society of Chemical Engineers, Japan, 23–25 March

[42] Lucas L L and Unterweger M P 2000 'Comprehensive review and critical evaluation of the half-life of tritium', *J. Res. Natl. Inst. Stand. Technol.*, **105**, 541–550

[43] Weart S R 1988 *Nuclear Fear: A History of Images* (Cambridge, MA: Harvard University Press)

[44] Dwight D Eisenhower Library, Abilene, Kansas USA (as of June 2004: http://www.eisenhower.utexas.edu/atoms.htm)

[45] Weart S R, op. cit. p 163

[46] Ewels C P *et al* 2003 *Physical Review Letters*, **91**(2), 11 July (APS paper reference: 025505)

[47] Nuttall J 1970 *Bomb Culture* (London: Paladin)

[48] National Security Archive, *US Freedom of Information Act at 35*, George Washington University, Washington, DC (as of June 2004: http://www.gwu.edu/~nsarchiv/)

[49] Hewitt G F and Collier J G, op. cit. p 146

[50] van der Pligt J 1992 *Nuclear Energy and the Public* (Oxford: Blackwell) p 3

[51] Hewitt G F and Collier J G, op. cit. p 164

[52] Hewitt G F and Collier J G, ibid. p 171

[53] Hewitt G F and Collier J G, ibid. p 175

[54] van der Pligt J, op. cit. p 8

[55] Hewitt G F and Collier J G, op. cit. p 181

[56] UKAEA, Press Release, 21 April 1998

[57] Nuclear Energy Agency 2000 *Nuclear Education And Training - Cause For Concern? A Summary Report* (Paris: Organisation for Economic Co-Operation and Development)

[58] Health and Safety Executive, *Nuclear Education in British Universities*, March 2002

[59] Coverdale T 2002 *Report of the Nuclear Skills Group* (London: Department of Trade and Industry), 5 December

[60] *Modern Apprenticeships—the Way to Work*. Report of the Modern Apprenticeships Advisory Committee, Sir John Cassels Chair, September 2001

[61] Conklin J, *Wicked Problems and Social Complexity*, Cognexus Institute, 2001–2003 (as of July 2004: http://cognexus.org/wpf/wickedproblems.pdf)

[62] Nirex UK, *A World of Science* (Oxfordshire: Harwell)

[63] Office for the London Convention 1972, International Maritime Organization, London, UK (as of June 2004: http://www.londonconvention.org)

[64] Parliamentary Office of Science and Technology, *Radioactive Waste Where Next?* London, November 1997

[65] Department for the Environment, Transport and the Regions and Nirex UK, *Radioactive Wastes in the UK—A Summary of the 1998 Inventory*, July 1999

[66] Department for Environment, Food and Rural Affairs and Nirex UK, *Radioactive Wastes in the UK: A Summary of the 2001 Inventory*, Nirex Report N/041, DEFRA/RAS/02.003

[67] UK-Centre for Economic and Environmental Development 1999 *UK National Consensus Conference on Radioactive Waste Management, Final Report*, Cambridge, UK

[68] House of Commons Science and Technology Select Committee Fourth Report, Session 2002–03 *Innovation in Nuclear Fission*

[69] RE Dunlap *et al* 1993 *Public Reactions to Nuclear Waste* (Durham, CA: Duke University Press) ch 4

[70] Slovic P *et al* 1979 'Rating the risks', *Environment*, 21, p 3

[71] World Nuclear Association, *Uranium Markets*, November 2003 (as of May 2004: http://www.world-nuclear.org/info/inf22.htm)

[72] Hon Mr Justice Parker 1978 *The Windscale Inquiry* (London: HMSO)

[73] House of Commons, *Hansard*, 13 April 1989, column 1132

[74] British Energy, *Replace Nuclear With Nuclear*, submission to the Performance and Innovation Unit's Energy Review, September 2001 (as of June 2004: http://www.british-energy.com/corporate/energy_review/energy_submission120901.pdf)

[75] Department for the Environment, Transport and the Regions, *An R&D Strategy for the Disposal of High-Level Radioactive Waste and Spent Nuclear Fuel*, DETR/RAS/99.016, London, October 1999

[76] Department for Environment, Food and Rural Affairs, *Managing Radioactive Waste Safely*, London, September 2001

[77] Nirex Information Office *World Leading Expertise in Radwaste Disposal*, Nirex UK, Harwell, Oxfordshire

[78] Nirex UK, *Policy Statement: Transparency*, February 1999 (as of June 2004: http://www.nirex.co.uk/foi/policies/transparency_policy.pdf)

[79] Department of Trade and Industry, *Managing the Nuclear Legacy*, London, July 2002

[80] House of Lords 1999 *Management of Nuclear Waste*, Select Committee on Science and Technology, HL Paper 41, London

[81] Royal Society 1998 *Management of Separated Plutonium*, London

[82] Mark J C, Taylor T, Eyster E, Maraman W and Wechsler J 1987 'Can terrorists build nuclear weapons?' in *Preventing Nuclear Terrorism*, Levevthal P and Alexander Y (eds) (Nuclear Control Institute)

[83] Levevthal P and Alexander Y (eds) *Preventing Nuclear Terrorism* 1987 (Nuclear Control Institute), p 434

[84] Evidence of Dr Derek Ockenden, Former BNFL Research Chemist, UK-Centre for Economic and Environmental Development, Consensus Conference Report, op. cit. pp 55–56

[85] Harris J, Science and Public Affairs, February 2000, pp19-21, Harris attributes the original phrase to Frank von Hippel

[86] Crowley K D, *Physics Today*, June 1997, pp 32–39

[87] Ahearne J F, *Physics Today* June 1997, pp 24–29

[88] Kastenberg W E and Gratton L J, *Physics Today*, June 1997, pp 41–46

[89] Posiva Oy, *Nuclear Waste Management of the Olkiluoto and Loviisa Power Plants, Annual Review*, 25 July 2002 (as of June 2004: http://www.posiva.fi/esitteet/YJH_engl_2001.pdf)

[90] Tang Y S and Saling J H 1990 *Radioactive Waste Management* (New York: Hemisphere), p 2

[91] Evans N and Hope C 1998 *Nuclear Power, Futures, Costs and Benefits* (Cambridge: Cambridge University Press), ch 6

PART II

NUCLEAR FISSION TECHNOLOGIES

Chapter 5

Water-cooled reactors

The UK is a special case where the issues that would be involved if the country were to embark on the construction of nuclear power plants are concerned. As described in chapter 3, among leading industrialized countries the United Kingdom has led the world in adopting competitive electricity markets. Concomitant with reforms to utility markets the UK has also shed the concept of 'national champions' in key industrial sectors. In almost every large-scale UK government procurement there is a genuine competitive process of bidding with relatively little emphasis given to issues of national industrial policy.

It seems certain that if the UK were to tender for the construction of new nuclear build, then the process of bidding would be competitive and domestic bidders would have little or no advantage. This means that if one examines the range of power plant options that would face a UK policy-maker keen to embark on a nuclear renaissance then one is led to consider several different technologies originating from several different firms and consortia from around the world.

The range of players in any such tender process for new UK nuclear build would inevitably include the nationalized nuclear company British Nuclear Fuels plc (BNFL). While BNFL is a long-standing British firm, it also owns Westinghouse Electric Company. This subsidiary has a long-standing US pedigree. The Westinghouse links to UK nuclear generation predate its acquisition by BNFL, as the most recent British civil power reactor, Sizewell B, is a Westinghouse-based pressurized light water reactor. The lack of any construction activity in the UK by those firms behind the large Advanced Gas-cooled Reactor programme of the 1970s and 1980s limits the British involvement in consortia pushing to deploy renaissance technologies in the UK.

As they wait for an anticipated British 'renaissance' nuclear power plant developers such as AECL from Canada, Framatome ANP from France and BNFL/Westinghouse have clearly learned the lessons of the past. At the

British Nuclear Industry Forum annual conference *Energy Choices 2003* the then Chairman of BNFL, Hugh Collum, outlined the main issue facing the future of the nuclear industry as 'perception, perception, perception'. At the same meeting, the director for Marketing, Analysis and Operations at AECL, Milton Caplan, declared that the key issue is 'economics, economics, economics'. I would synthesize their views by saying that the two key issues are economics and perception.

Both these firms have confronted the modern realities of nuclear power by developing modular designs with improved economics. Both designs build upon proven technologies and combine these experiences with a desire to simplify and modularize wherever possible.

In each case it is likely that reactors would be built in fleets of identical units, so as to develop savings based upon organizational learning and economies of scale, although as we shall discuss, there may be security of supply benefits in having more than one reactor design within a nuclear power portfolio. To match the construction benefits arising from a large-scale national programme, both firms also seek streamlined, even international, licensing so as to reduce the need for delay and duplication in the regulatory processes. Potential providers of a nuclear renaissance regard project finance and regulatory uncertainty as the key economic risks facing any programme of new nuclear build.

In the following sections we shall examine the attributes of some water-cooled reactor technologies that could be deployed in the first wave of a UK nuclear renaissance. The examples chosen are not the only possibilities, but they do, between them, capture most of the technological issues associated with new nuclear build in the short-term.

II.5.1   Westinghouse BNFL—advanced passive series

Westinghouse Electric Company LLC, a BNFL Group company, proposes to launch the nuclear renaissance in western Europe and North America with its advanced passive (AP) series of pressurized light water reactors (PWRs). Westinghouse's AP system is available in two variants, a 600 MWe (megawatt-electric) unit known as the AP600 and a 1000 MWe unit known as the AP1000.

These power generation systems build upon the long experience of Westinghouse Electric in pressurized light water reactor systems. This experience dates as far back as the early 1950s when Westinghouse partnered with Argonne National Laboratory in the production of the Submarine Test Reactor (1953) and in the development of the world's first nuclear powered submarine the USS *Nautilus* [1]. The *Nautilus* served with the US Navy from 1954 until 1980. US PWR technology was first devoted to civilian power production at Shippingport on the Ohio River near Pittsburgh. This was the first civilian power reactor in the United States and the first civilian light water reactor anywhere in the world.

Figure II.5.1 illustrates the overall design of the AP600 and AP1000 systems. It is noteworthy that there is no difference in the footprint of the two designs, the necessary extra volume of the AP1000 being accommodated by a greater height for the containment building. This extra height makes possible the use of longer fuel assemblies. (AP1000 has a 14 foot tall reactor core as opposed to the 12 foot long core used in the AP600 [3].) In addition to the larger fuel assemblies, the AP1000 has a larger number of fuel assemblies and larger steam generators than the AP600 design.

The larger containment of the AP1000 provides additional free volume. This additional free volume, and the use of different materials in the vessel shell, provide for extra resilience in the event of an accident-related pressure build up within the containment [3].

Westinghouse [2] emphasize the following attributes of the AP1000 that represent an improvement over previous pressurized light water reactor systems.

II.5.1.1   Safety

As the name suggests the Westinghouse Advanced Passive series of power reactors make use of passive safety features; and safety-critical pumps, fans, diesel generators and other rotating machinery are all avoided [2]. The design relies substantially upon natural attributes of gravity, convection and the properties of compressed gases to ensure safety in the case of system disruption. In addition, as far as possible, valves are fail-safe. That is, in the event of a power failure the valves would default to their safer settings. The use of passive systems not only improves the simplicity and the safety of the reactor, it also reduces its cost. Figure II.5.2 illustrates that the AP designs involve significantly fewer engineered components than previous PWR designs.

Extensive work has been done in ensuring the resilience of both AP600 and AP1000 in the face of moderate loss of coolant accidents. Even for an 8 inch (200 mm) break in the vessel injection lines of the reactor coolant system there is predicted to be no possibility that the reactor core could become uncovered in either the AP600 or AP1000 designs [3].

Regarding British regulator attitudes to the safety features of water-cooled reactors, such as the AP1000, it is interesting to note that the only light water civil power reactor in the UK at Sizewell B was required by regulators to incorporate extra safety systems including an Emergency Boration System that could inject a highly concentrated boron solution into the core following any failure of the control rods to drop into the core [4]. These requirements went beyond the standard US safety systems for PWR reactors. The requirements of the British safety regulator added significantly to the

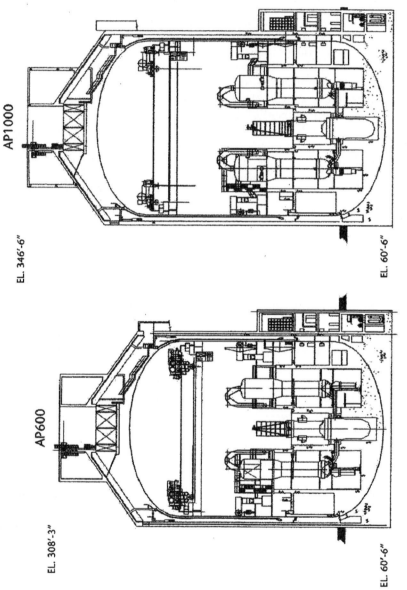

Figure II.5.1. Cross-sectional view of the Westinghouse AP600 and AP1000 reactor containments (source: BNFL).

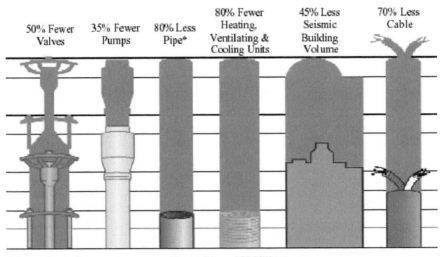

| 50% Fewer Valves | 35% Fewer Pumps | 80% Less Pipe* | 80% Fewer Heating, Ventilating & Cooling Units | 45% Less Seismic Building Volume | 70% Less Cable |

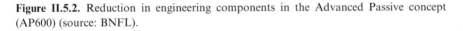

* Safety Grade Compared to a conventional, 2-loop 600 MWe plant

Figure II.5.2. Reduction in engineering components in the Advanced Passive concept (AP600) (source: BNFL).

cost and complexity of the Sizewell B system. The attitude of the British safety regulator to the Advanced Passive concept remains to be seen.

The AP series involve containment designs based upon standards established over a long history of civilian light water reactor operation. Robust containments, such as the type adopted for the AP series, are now required by US safety regulators.

Westinghouse has undertaken extensive probabilistic risk assessment of the AP series designs. These analyses have predicted that the frequency of core damage is two orders of magnitude lower than the requirements of electricity utilities and regulators (1×10^{-5} events per year).

II.5.1.2 Modularization

The Westinghouse AP series are modular in two respects: structural elements and system blocks [2]. The structural modules consist of prefabricated assemblies combining walls, pipe-work and loomed cabling. These structural blocks are transported to the construction site and incorporated into the build as a single large element. The AP designs involve more than 300 such modular elements each of which would be prepared away from the construction site and in parallel with the main build programme [2]. Modularization has several benefits including easier economies of scale for fleet construction, improved quality control and, above all, accelerated construction schedules.

II.5.1.3   Rapid construction

Through the use of modularity and vastly simplified engineering (involving fewer engineered elements) the AP series reactors have been designed to allow for extremely rapid construction. Westinghouse confidently reports that complete construction can be implemented on budget within a three-year envelope. Reliable and fast construction is vital if the nuclear renaissance is to become a reality. Nuclear power plant construction in the first wave (say up until 1990) suffered from a frequently well-deserved reputation for cost overruns and delays. Often the major delays related to regulatory and licensing issues, but all too frequently the delays resulted from engineering difficulties. In the case of AP1000 it seems likely that the engineering will indeed be more reliable. Primarily this is as a consequence of the plant being a fully developed design rather than part of the craft tradition of 'learning by doing' seen in some nuclear projects. Westinghouse has also emphasized the use of tried and tested technologies and avoided the temptation to regard the first AP series plants as research and development projects.

II.5.1.4   Improved economics

The combined attributes of passive safety, simplified design, modular construction and accelerated construction, all lead to reduced capital cost and minimized economic risk. These are two key requirements of any system that its proponents expect to be at the forefront of any nuclear renaissance. AP600 incorporated these concerns, but Westinghouse recognized that the estimated cost of AP600 electricity of 4.1–4.6¢/kWh was likely to be uncompetitive in the liberalized US electricity market. Westinghouse therefore embarked on the development of the AP1000 which, via economies of scale, was expected to achieve final electricity costs of about 3.0–3.5¢/kWh [3]. More precisely, the Westinghouse AP proposal combines capital costs estimated at US $1150 for the third twin AP1000 unit in a fleet with fully-costed analysis of fuel purchase, operations and maintenance, waste disposal and decommissioning. With an anticipated plant lifetime of 60 years and an operating availability of 93%, the overall generation cost for the AP1000 has been described by Westinghouse as being $36 per MWh (3.6¢/kWh), consistent with lower level predictions for combined cycle gas turbine–natural gas electricity generation at current natural gas prices [2]. A 2002 Westinghouse analysis of the cost of AP1000 electricity is reproduced in table II.5.1.

On 6 January 2000 the smaller AP600 reactor received full design certification from the US Nuclear Regulatory Commission. Westinghouse had first filed papers with the NRC relating to the AP600 in June 1992. The NRC decision of January 2000 represented a key step towards regulatory approval of Westinghouse's Advanced Passive philosophy. The NRC design certification paved the way for the AP600 to be offered to US utilities

Table II.5.1. The Westinghouse Electric predicted cost for AP1000, assuming the third twin unit in an AP1000 fleet operating at 93% availability for 60 years (source: C K Paulson/BNFL [2]).

Cost component	US $
Capital per kWe	1150
Fuel per MWh	5
Operations and maintenance per MWh	5
Disposal per MWh	1
Total generation cost per MWh	36

as a candidate for new nuclear build. BNFL must surely have been pleased that their March 1999 acquisition of Westinghouse Electric did indeed provide the British company with a reactor design that passed regulatory approval. The AP1000 is expected to receive its NRC design certification in 2005.

II.5.1.5   AP1000 and MOX

As noted previously, this author regards two well-established UK policy axioms as being largely unhelpful. The first is the attempt to distinguish between legacy wastes and wastes arising from future nuclear power plant developments. The second is the apparent need to decide whether separated civil plutonium is a waste or an asset. The dominant view within the British nuclear industry is that plutonium is indeed an asset and that it should be thought of as a fuel—probably as part of MOX fuel. If, however, plutonium is declared a waste, then it is not the case that MOX strategies should immediately be dismissed. MOX has some merits as a plutonium waste management strategy. Any declaration that plutonium is a legacy waste, or alternatively that it is a fuel for a new fleet of power plants, is in this author's opinion unnecessary and overly restrictive. The issues are too complex to benefit from such compartmentalization. What seems to be needed is a holistic examination of all the issues, before a plutonium management strategy is adopted. Such a holistic assessment might in fact result in powerful arguments in favour of MOX-based plutonium management. AP1000 would be a technology well placed to contribute, given its designed-in ability to operate with 100% MOX fuel. With Britain and plutonium we do not start from where we would like to be, but from where we actually are.

II.5.1.6   Criticisms of the AP1000

Much criticism of the Westinghouse AP series relates to the old rivalries among advocates of different reactor systems such as gas-cooled reactors, boiling water reactors and heavy water reactors. While the highest nuclear fuel burn-ups have been achieved in PWR systems, it is noteworthy that other technologies can extract untapped energy from spent LWR fuel without the need for full reprocessing (e.g. via the DUPIC fuel cycle discussed in the context of the AECL ACR). Given continued low uranium prices (figure I.4.5), however, such considerations would seem to be of little current concern. The greatest economic consequence of improved fuel burn-up could well be in considerations of waste minimization, rather than fuel purchase costs, but even so burn-up alone is unlikely to be a deciding factor for the fuel cycle.

Probably the most stinging criticism of the BNFL/Westinghouse proposal is in fact a criticism of BNFL/Westinghouse as a company, rather than of the AP1000 design. The criticism runs that it is more than 20 years since Westinghouse Electric constructed any civil power reactor anywhere. The implication of such comments is that Westinghouse has lost its core capabilities in this area and that much important tacit knowledge has been lost to the firm over the past two decades. The firm is conscious of these difficulties and responds by pointing out that the central technologies of AP1000 are 'real world' proven, not only from civil power plant work but also in its ongoing involvement in US Navy propulsion reactor systems. W E Cummins and R Mayson also report that all major components of both the AP600 and the AP1000 have been proven in operating reactors under similar flow, temperature and pressure conditions, except for the AP1000 reactor coolant pump [3]. (It is reported to be a modest extension of proven pump designs.) In addition BNFL/Westinghouse adopts a consortium approach to the development of the AP series reactors. Knowledge is held among a range of companies and such knowledge is in fact far more resilient than the critics claim. One specific aspect of the consortium approach is the acquisition by BNFL of the nuclear business of ABB, which brings to the group recent experience with commercial nuclear power plant development in the Far East.

As will become even clearer in the overview of the competing designs that follows, firms aiming seriously to lead the nuclear renaissance must have low and reliable costs and short construction times.

II.5.2   Atomic Energy Canada Limited—Advanced CANDU Reactor

The Atomic Energy Canada Ltd (AECL) offering for initial deployment in the nuclear renaissance is the Advanced CANadian Deuterium Uranium (CANDU) Reactor, or ACR. This light water-cooled, heavy water moderated thermal reactor is designed to produce 700 MWe. Aware that since the ending of the AGR programme the UK lacks a truly home-grown and preferred reactor design, AECL is ready to pursue the possibility of orders in any British renaissance.

Appealing to the strong cultural ties between the two countries, AECL emphasizes over one hundred years of collaboration between British and Canadian scientists and engineers in nuclear matters. AECL point out that Dr Ernest Rutherford moved from the Cavendish Laboratory of Cambridge University to become Professor of Physics at McGill University in Montreal in 1898.

Of more direct relevance to a possible nuclear renaissance in the 21st century is the fact that in November 2001 AECL and the nuclear utility British Energy agreed to collaborate on 'work to assess the feasibility of CANDU technology as a potential nuclear power station option in the UK' [5]. At the heart of this relationship was to have been British Energy's experience gained in operation of the eight-unit Bruce Nuclear Power Development (BNPD) power plant in Ontario. British Energy's Canadian experience came from its lease of the BNDP plants. The lease had the potential to be highly profitable for British Energy, while also giving the British company experience with the CANDU system. Unfortunately the Canadian deal required substantial investments of cash, both to maintain operation of the working BNPD units and to bring back on-line two units that had been shut down for a long period. Cash, however, was something in very short supply at British Energy at that time. Back home in the UK the NETA electricity market had resulted in unexpectedly low wholesale electricity prices. As a generator, British Energy had pursued a strategy of international acquisitions in nuclear generation. Its UK electricity competitors, however, had pursued a strategy of vertical integration combining their generating power with retailing operations. When the generation market was squeezed, these competitors were able to compensate their shortfalls with continuing success in the retail arena. The consequence of the new UK market and British Energy's business planning was that in 2002 the company found itself on the edge of bankruptcy and requiring emergency support from the UK government.

As part of the government rescue package, the British Treasury required that British Energy divest itself of its Canadian acquisitions. The forced sale of the British Energy Canadian assets probably yielded a poor return for Britain and a very good deal for the purchasers, a Canadian consortium led by the uranium producers Cameco.

Whatever the future of the strategic alliance between AECL and British Energy, it seems clear that AECL still hopes to persuade UK policy makers and utilities of the merits of its ACR design. The key attributes of the ACR proposal are as follows.

II.5.2.1 Fuel attributes

In comparison with pressurized light water reactors the CANDU system has several fuel management benefits. The reactor design permits on-power refuelling, increasing the availability of power to the distribution grid.

With its heavy water moderator the ACR system has improved neutron economics over its LWR competitors. The higher burn-ups achievable in the ACR system are such that, following refabrication, it is even possible to re-use fuel that has reached the end of its useful life in PWR systems. This refabrication is not reprocessing in the conventional sense, as there is no need to separate out any of the chemical components of the used PWR fuel. Furthermore it is a dry process known as 'DUPIC' (Direct Use of spent Pressurized water reactor fuel In CANDU), which has been explored in depth in recent years by AECL in its Korean operations [6]. DUPIC is controversial, however. While it is indeed a far simpler process than reprocessing, it is nevertheless a technically complex process. A key difficulty is that it involves the manipulation of spent PWR fuel. As such, substantial handling facilities are required for extremely radioactive materials. Such hot-cells are expensive and complex technologies in their own right and their costs would surely limit the economic viability of the DUPIC fuel cycle. Furthermore, the DUPIC approach does not add enormously to the burn-up achieved in the nuclear fuel cycle, being widely regarded as giving an improvement of only about 10% in fuel burn-up. Lastly the DUPIC fuel cycle relies on a nuclear power programme employing both LWRs and CANDU power plants. Such a situation is rather unusual among nuclear power developments around the world. South Korea is a notable exception in that regard.

Fuel efficiency is a key improvement of the ACR over earlier CANDU designs. The ACR uses slightly enriched uranium (SEU) fuel rather than the natural uranium employed in earlier CANDU designs. (Optimum ACR performance is expected using SEU of approximately 2.1% uranium-235 by weight [7].) The use of natural uranium in the original CANDU reactors was a key benefit to Canada in its early development of nuclear power. Canada is one of the world's leading suppliers of uranium and supplies of the metal were always likely to be plentiful for nuclear power in Canada. The use of natural uranium in a water-moderated reactor requires the use of heavy water, as it has a much lower thermal neutron absorption cross section than light water [8]. Ideally reactor designers seek neutron moderation without the complication and difficulty of maintaining criticality in a reactor system where a significant proportion of the neutrons are lost via absorption in the moderator. Ideally neutron absorption should be a controlled property dominated by the control rods.

By adopting a fuel cycle based upon natural uranium Canada was able to develop nuclear power without the need for a costly fuel enrichment infrastructure. The use of natural uranium, however, and the original CANDU use of heavy water as both coolant and moderator, resulted in lower operating temperatures than PWR competitors and hence lower thermodynamic efficiencies [9]. As fuel enrichment technology is now widely available and cost effective, the designers of the ACR have been able to adopt the use of SEU fuel and light water-cooling with significant improvements in reactor thermodynamic efficiency and fuel burn-up properties. Another improvement arising from the move to SEU and greater fuel burn-up is the concomitant reduction of spent fuel volume produced per MW of electricity generated. These improvements all contribute to the improved economics of ACR over its CANDU predecessors.

II.5.2.2   Plutonium disposition

A key attribute of the ACR is its capacity for the highly efficient utilization of mixed-oxide or MOX fuels. The ACR design is such that it readily accommodates up to 100% of its fuel loading in the form of MOX fuel bundles [11]. As discussed in chapter 4, MOX fuel combines uranium and plutonium oxides and utilizes the thermal neutron fissile properties of the uranium-235 and plutonium-239 isotopes. While international uranium prices remain low it is not easy to justify the use of MOX fuels on economic grounds. The primary justification for the use of MOX fuels, therefore, is one of plutonium disposition. Because of its proliferation risks, separated plutonium requires special management, which can be both costly and burdensome. Several countries, most notably the UK and Russia, have significant stockpiles of plutonium, and MOX represents one method of disposal (see chapters 4 and 7). MOX fuels are most usually discussed in the context of light water reactor systems where significant operational experience has been gained. True disposal of plutonium in PWR MOX requires multiple iterations of the plutonium around a recycling-based nuclear fuel cycle—a process that is both costly and complex. By contrast, the use of MOX fuels in CANDU reactors such as the ACR ensures significant plutonium burn-up in a single path. Spent MOX fuel emerging from an ACR may appropriately be regarded as high-level waste with any remaining plutonium protected to a spent-fuel standard that will remain proliferation-resistant in perpetuity. The proliferation resistance of LWR MOX also satisfies the spent-fuel standard, but the very long-term evolution of the isotopes in spent LWR MOX is less proliferation-resistant than would be expected from ACR MOX.

II.5.2.3   International construction experience

AECL is proud of its long-standing experience in nuclear power plant construction. Unlike most other firms with histories in nuclear power plant construction, AECL has a continuous and ongoing record in nuclear reactor system construction. Ten CANDU-6 700 MWe reactors have been constructed by AECL and its partners with in-service dates ranging from February 1983 to July 2003. The CANDU-6 systems each have a stand-alone single unit reactor. The CANDU-6 series of reactors followed and overlapped with the construction of the 900 MWe class of CANDU reactors. In total, twelve 900 MWe class reactors were constructed by AECL and

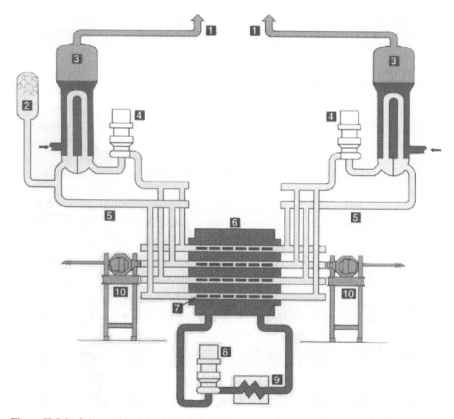

Figure II.5.3. Schematic of the ACR 700 MWe reactor showing the calandria (6), the light water-cooling circuit (5) and the heavy water moderator pump and heat exchanger (8, 9). Refuelling can be performed on-power using machines (10). The reactor uses essentially identical steam generators (3), pressurizers (2) and heat transport pumps (4) as a PWR. The steam pipes (1) take the steam to the turbines—see figure II.5.6 (source: AECL).

partners in Canada at two sites, Bruce and Darlington in Ontario. These larger CANDU series of reactors had in-service start dates ranging from September 1977 to June 1993 [1]. The Bruce site has eight CANDU reactors, four of which are rated at 825 MWe and four of which are rated at 915 MWe. For more than five years two of the 825 MWe reactors were shut down (units 3 and 4) until their return to operations in late 2003 following the forced sale of Bruce Power by British Energy [12]. AECL has a long tradition of continued progress in reactor design and construction. It is now a plausible and natural step for the company to make the move from CANDU-6 to ACR.

In addition to constructing the CANDU series of reactors worldwide, AECL has a long and ongoing track record in the construction of research reactors for neutron beam studies and isotope production. AECL's current series of research reactors are known as MAPLE and three reactors with MAPLE cores were constructed by AECL and its partners during the 1990s [13].

II.5.2.4 Modularity

The ACR design employs several innovations in construction focused on reducing capital costs and improving initial build quality. AECL has moved aggressively to employ three-dimensional computer aided drafting and design (3D CADD). This has not only assisted the plant designers, but it also facilitates the work of those involved in local procurement and equipment specification and it further supports the work of those involved in plant commissioning and operation. Perhaps most importantly for minimized construction costs the use of 3D CADD allows for complex installation planning, including sequencing and orienting the installation of a series of prefabricated elements of widely varying shapes and sizes.

Modularized construction has not only been facilitated through the use of 3D CADD, but it has also been dramatically assisted by the use of open top construction and heavy-lift cranes. AECL has successfully employed the open top method in its construction of two CANDU-6 units at Qinshan in China, building upon earlier experience gained in the construction of three CANDU-6 units at Wolsong in Korea. In the open top method the dome of the reactor containment is one of the last elements to be built. This allows for a sizeable hole to remain in the roof of the reactor building during most of the construction period. Such a gap is perfect for allowing major items of equipment such as steam generators to be lifted by heavy-lift crane in through the roof of the reactor building. This allows a significant saving of time and effort as a steam generator can be installed in one or two days rather than the two weeks typically required if conventional horizontal access techniques are used [6]. See figure II.5.4.

II.5.2.5 Lack of a single pressure vessel

The traditional AECL CANDU design was characterized by the use of natural uranium fuel and heavy water as both coolant and moderator. If such a reactor were to use a conventional pressure vessel as in a light water reactor, then because deuterons in heavy water are a less efficient neutron moderator than protons in light water, more collisions and hence a larger volume of moderator would be required [14]. This would result in a very large reactor vessel indeed. Siemens has in fact developed a heavy water reactor design of exactly that type and one such unit, 340 MWe, has operated successfully at Atucha in Argentina (it became the first operational power reactor in Latin America in 1974). Its sister pressurized heavy

Visions of the renaissance I: UK 2015

In this box we consider just one possible future path for nuclear power in the UK. It is a relatively unconventional prediction and as such should not be regarded as a description of the future. It is merely a possible scenario intended to provoke thinking concerning the future of nuclear power in the UK.

In this scenario the UK has changed its energy policy to state clearly that approximately 25% of national electrical power should be provided by nuclear energy. In effect, the government has adopted the British Energy recommendation of 2001 that the UK should 'replace nuclear with nuclear' [10].

By 2015 the ability to invoke the DUPIC process to extend the life of LWR fuel has opened up a powerful new opportunity within the first wave of the nuclear renaissance. In the early years of the century many regarded the AP1000, EPR and ACR designs as simply competitors attempting to capture orders for new nuclear power stations. The DUPIC process, however, favoured the possibility that both PWRs and ACR might be deployed simultaneously. Such a strategy has the following possible advantages.

*   *It fosters a spirit of competition between constructors each aiming to demonstrate that their technology can be assembled to the higher quality, timeliness and cost efficiency. Lessons must, however, be learned from the UK experience with the AGR programme in which a desire for competition between constructors led to wasteful duplication of effort and a damaging lack of standardization. The nature of the competition between the PWR developers and AECL would therefore need to be clear and simple.*
*   *It reduces the technical risk to security of electricity supplies in the event that any one technology is revealed to suffer from a design flaw requiring rectification across its fleet. (Such a scenario occurred with the AGR gas recirculators in 2002.) If the nuclear power portfolio for the country includes a variety of technologies (say two, or three at most), rather than a single design, then disruption caused to the national electricity system by the withdrawal of any single reactor type would be limited.*
*   *It provides the possibility, via the DUPIC process, of using refabricated PWR fuel. The ACR system permits the efficient utilization of nuclear fuel without the need for costly and complex reprocessing.*

By 2015 the UK has shut down all of its Magnox generating plants and has withdrawn from commercial nuclear fuel reprocessing with the closure of the THORP plant at Sellafield. Domestic nuclear power is now provided by three distinct technologies: the gas-cooled reactors (AGRs), now approaching the end of their operational lives; the pressurized light water reactors (Sizewell B and several new PWR plants) and a new ACR station comprising two heavy water moderated reactors. A DUPIC plant has been constructed at Sellafield and there are plans to construct further PWR and ACR plants to replace the capacity of the AGR stations as they retire.

Thus far the PWRs have been running on MOX fuel as part of a strategy of legacy plutonium management. Investigations are under way regarding the possible use of MOX fuels in the new ACR plant. No new plutonium separation is occurring in the UK and it is intended that in future spent PWR fuel will be declared waste or be passed into the DUPIC process. However, it will be some years until the ACR plant runs routinely with DUPIC fuel. By 2015 the government has already declared its intention that used DUPIC and MOX fuels will be declared 'waste'.

water reactor Atucha II (also a Siemens design) was never completed, construction having been abandoned in 1995 when it was more than 80% complete and more than 525 tonnes of heavy water had been brought to the reactor [15].

The AECL response to the difficulty of neutron moderation in a moderately sized reactor was both imaginative and innovative—the calandria (see figure II.5.5). The calandria avoids the use of a single pressure vessel and this innovation has become the key technological characteristic of all CANDU reactors, including, in modified form, the ACR.

The calandria concept consists of a large horizontal cylindrical block with hundreds of horizontal fuel channels of circular cross section running the length of it with each channel open at both ends. The calandria is not a pressure vessel and it is not subjected to the high pressures that are required in order to circulate liquid reactor coolants. Rather the calandria must hold the heavy water moderator without leakage and permit its gentle recirculation. The circulation of high pressure coolant is made possible by pressure tubes that run through the calandria. It is these several hundred pressure tubes that serve the function of the pressure vessel in light water reactors. The CANDU fuel assemblies run in the calandria channels within the coolant pressure tubes. Refuelling, which can be performed on-power, is achieved using two machines, one at each end of the calandria. One machine pushes in a fresh fuel bundle while the other receives the spent fuel bundle pushed out of the other side.

Figure II.5.4. AECL use of open top construction techniques in Qinshan, China (source: AECL).

Figure II.5.5. Calandria under construction in Korea for the Wolsong unit 4 CANDU-6 plant (source: AECL).

The ACR design retains the calandria concept, but because of the use of slightly enriched uranium fuel (and a light water-coolant circuit) the calandria can be more compact than that in the CANDU-6 and CANDU 900 MWe series. The compactness of the calandria further reduces construction costs and eases construction. A schematic of the ACR design is shown in figure II.5.6.

II.5.2.6 Economics

The driving principle behind the development of the ACR design has been that nuclear power must compete economically in deregulated electricity markets with other forms of base load generation such as natural gas fuelled combined cycle gas turbine (CCGT) plants. That is, in Europe the lifetime average cost of generation must compete with likely CCGT costs of US $30–45 per MWh. This cost range is a benchmark that AECL have addressed with particular enthusiasm. Recognizing that historically the capital cost of plant has been problematically high, several innovations of the ACR over previous CANDU designs push in the direction of reduced construction costs. These developments include:

- the use of light water-coolant rather than expensive heavy water,
- a smaller calandria made possible by the use of slightly enriched uranium fuel,
- rapid open-top construction techniques using heavy lift cranes and
- widest possible economies of scale as 75% of the internal components of the ACR are the same as PWR technology [16].

These innovations combined with the ongoing track record of the CANDU-6 series have resulted in AECL declaring that it is willing to construct ACR systems at fixed cost. That the constructor is willing to take on board the risk of cost escalation must be a welcome message for energy policy makers with memories of nuclear power plants over budget and behind schedule. Perhaps even worse for the industry than its high capital costs has been the uncertainty and the economic risk borne by governments and utilities during construction. That a constructor is willing to assume this risk must be a welcome development for policy makers and utilities alike.

II.5.2.7 CANDU: a civilian design for civilian needs

It is widely known that during World War II many British nuclear scientists travelled to Los Alamos in New Mexico, USA, to assist with the Manhattan project. What is less widely known is that many of the researchers of the University of Cambridge Cavendish Laboratory moved to Montreal to work with the National Research Council of Canada on heavy water research. Of the approximately 150 staff to relocate in 1942 one was Professor John

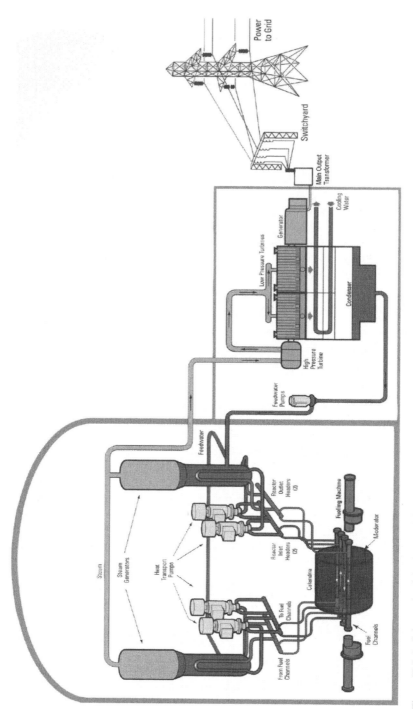

Figure II.5.6. Schematic of the AECL Advanced CANDU Reactor (ACR) showing reactor containment and connection to the power grid (source: AECL).

Cockcroft (see chapter 3, section I.3.6). Professor Cockcroft and his Cavendish colleague Ernest Walton would later be awarded the Nobel Prize in physics for their early 1930s work in Cambridge, England.

Under Cockcroft's leadership the heavy water researchers soon moved to a large new laboratory at Chalk River in Ontario which still today is the leading Canadian nuclear research laboratory [17]. As early as 1952 Canada refocused its nuclear research efforts towards civilian needs with the creation of AECL. The CANDU design was developed during the decade that followed with a 24 MWe demonstration reactor being completed in 1962 at Rolphton, Ontario. It is noteworthy that, unlike its international contemporaries the US light water reactors and the British Magnox system, the CANDU approach never received synergetic benefits from parallel military programmes, such as those in naval propulsion or nuclear weapons development. The CANDU design may be regarded legitimately in its entirety as having been a wholly civilian design for wholly civilian needs. With its military aerospace links, the claim cannot be made for the gas-fired electricity generation technology—the combined-cycle gas turbine.

The peaceful credentials of CANDU were, however, badly tarnished in the early 1970s by the illicit actions of India, the operator of two CANDU units. Its Rajasthan Atomic Power Station (RAPS) facility developed what was, in effect, India's first nuclear weapon. The 1974 Indian 'Peaceful Nuclear Explosion' or PNE at Pokhran was in reality both a nuclear weapons test and a demonstration to the world that India considered itself to be a great power member of the nuclear club. As the international partners in the Indian civil nuclear programme AECL found themselves implicated in the clandestine military activity. It is widely believed that the plutonium used in the Indian PNE had been extracted from fuel used in the CANDU reactor supplied by AECL. AECL and Canadian nuclear scientists were for many years stigmatized by the other nuclear powers as having been implicated in nuclear weapons proliferation. However, no real sanction was ever applied to the United States, which had supplied the heavy water to India necessary for the operation of the RAPS plant. Following the PNE, AECL dissociated itself from the Indian CANDU programme, which continues to this day on a separate path of development.

The ACR, with its use of slightly enriched uranium fuel, has, when combined with a once through fuel cycle and strong international safeguards, the potential to be a thoroughly proliferation resistant technology. It is both unfortunate and inevitable that the Indian PNE story will always lurk in the background when CANDU and proliferation are considered.

II.5.2.8 ACR and the weaknesses of CANDU

By abandoning the outdated emphasis on natural uranium, AECL has, in its ACR design, been able to fix some of the clear shortcomings of the traditional CANDU approach. Although the use of the calandria eliminated the need for a very large pressure vessel, the combination of natural uranium fuel and heavy water moderation inevitably implied that the core of traditional CANDU reactors would have to be very large. By moving to slightly enriched fuel, the calandria in the ACR design is significantly smaller than in previous CANDU reactors. This eases construction and reduces capital costs.

Another long-standing weakness of the traditional CANDU reactor has been tritium production and leakage. Tritium (half-life 12.3 years and a beta emitter) is a highly radioactive heavy isotope of hydrogen inevitably produced in heavy water reactors. Although the neutron adsorption cross section of deuterons (the hydrogen atom nuclei in heavy water) is much lower than the equivalent cross section in light water, it is not zero. When deuterium atoms capture a neutron they become tritium atoms. The tritium is retained within heavy water molecules as a radioactive liquid.

The traditional CANDU design has a lengthy high-pressure heavy water-cooling circuit. The heavy water (containing the tritium produced by neutron capture) is held liquid while at high temperature by the high pressure within the cooling circuit. In the traditional CANDU design if there were to be a leak from the primary cooling circuit it is likely that heavy water steam (made radioactive owing to its tritium component) would be emitted. The move towards light water-cooling in the ACR is a welcome move in favour of worker safety as it eliminates the possibility of significantly radioactive steam leaks. Furthermore, it reduces annual losses of expensive heavy water, which has been estimated for CANDU systems to be approximately 2% of the heavy water inventory of the reactor [8].

As noted previously, the use of a light water-coolant also reduces significantly the volume of expensive heavy water required and hence the capital cost of the ACR when compared with previous CANDU systems.

As regards safety, CANDU designs have been criticized for their performance in loss of coolant accidents or LOCAs. In particular the horizontal channels of the calandria are such that, if steam voids are formed, the water phase will settle in the lower part of the channel. The steam in the upper half of the channel will have far diminished cooling properties [18]. In addition Hewitt and Collier point out that the CANDU design is vulnerable to the related concern of coolant stagnation, that is, the very remote possibility that liquid coolant becomes stationary within the cooling channel. In that region the reactor quickly heats and steam voids may form. As a result of concerns about LOCA situations, traditional AECL CANDU designs (and the ACR) employ a back-up emergency core cooling system. The function of the CANDU emergency core cooling (ECC) system is similar to that used in conventional US light water reactors. The purpose of the ECC is to maintain core cooling in the event of a small leak from the primary cooling circuit. As the moderator and the cooling system are largely functionally separate in the CANDU design, the emergency cooling system does not involve the extra complexity of emergency neutron absorption. The use of gadolinium as an emergency reactor shutdown mechanism is separate from the CANDU cooling circuit and this will be discussed below. The high level of separation of these systems in CANDU designs is quite different from the situation in LWRs.

Britain's experience with heavy water moderated reactors during the 1970s and 1980s raises some possible issues regarding the operation of AECL CANDU reactors. While generally it operated quite successfully, the Steam Generating Heavy Water Reactor (SGHWR) at Winfrith in Dorset experienced problems of leaks from its calandria pressure tubes. The resulting repairs were difficult and caused plant outages and reduced the plant's overall load factor. The SGHWR plant also suffered from high levels of ambient tritium. This caused difficulties for reactor operations (e.g. the management of worker dose exposure) and it has recently been causing some difficulties for plant decommissioning. While it would be unfair to link the historical difficulties of the SGHWR to AECL and its ACR, it is inevitable that any British assessment of ACR would be shaped by the Winfrith experience.

A significant criticism of the traditional CANDU design is that it suffers from a positive void coefficient. That is, if gaps arise in the flow of heavy water-coolant then the reactivity increases. This is because the heavy water in the cooling channels normally has two neutronic consequences: first it helps moderate the neutron energies and second it absorbs neutrons. In the event of a steam void forming, the amount of neutron moderation is only slightly affected (as the main moderation is provided by the separate heavy water moderator) but the level of neutron absorption is significantly reduced. Hewitt and Collier report that within the first second of a LOCA and a void forming the fuel power can double [18]. This concern, together with the adverse impact on reactor cooling of steam voids in the channels, further motivates the need for multiple reactor shutdown systems. The CANDU reactor employs two separate shutdown systems in addition to the ECC discussed earlier. The first such system involves the gravity drop of solid mechanical cadmium rods into the reactor assembly. These cadmium rods soak up neutrons and immediately halt the nuclear chain reaction. It is important to note that because of the use of a calandria, rather than a single pressure vessel, these emergency shutdown rods do not need to operate under high temperature and pressure conditions. This can be expected to improve their reliability compared with their light water reactor analogues [18]. The second shutdown system in the CANDU design involves the use of a concentrated gadolinium-based neutron poison being injected into the low-pressure moderator circuit.

The separate moderator circuit is a source of extra comfort in the CANDU design as it inevitably absorbs some heat and can act to slow temperature fluctuations in the calandria [18]. The resulting long lead times for operator action are a welcome addition to the various safety benefits of the CANDU system.

Despite the admirable safety attributes of the traditional CANDU design and its outstanding 35-year safety record, the positive void coefficient is a particularly problematic issue. The problem in some ways goes beyond the merely technical issues to include significant issues of perception and human response. This is because the positive void coefficient (and a negative fuel coefficient) contributed to the problematic positive power coefficient at the heart of the Chernobyl RBMK reactor accident in 1986. The behaviour of the RBMK design at low power was such that the reactor responses to small changes in power could be dangerously uncontrollable. The positive void coefficient in the RBMK design was a significant contributor to the world's worst nuclear accident. The fact that the design and operation of the CANDU reactor is completely different (and inherently more stable than the RBMK) does not alter the fact that since Chernobyl, wherever possible, reactor designers have sought to avoid positive void coefficients (and negative fuel coefficients). It is welcome, and unsurprising therefore, that AECL report that the ACR has been designed in such a way that it will have a negative void coefficient of reactivity [19].

In addition AECL reports that the ACR will surpass the safety features of the CANDU-6 system in the following other respects:

- limited challenge to containment barriers, even in the event of a severe accident,
- highly stable core design,
- feeder corrosion minimized through materials selection,
- large operating margins and
- long lead-times for operator intervention.

With its enhanced safety attributes and its strengths in fuel management and economics it seems that the Advanced CANDU Reactor from Atomic Energy Canada Ltd really is a contender for deployment in the first phase of the nuclear renaissance.

II.5.3   European Pressurized Water Reactor (EPR)

The two competing firms BNFL/Westinghouse and AECL Ltd are not the only firms well placed to propose a new nuclear power plant for the United Kingdom. A European competitor exists to challenge the North American players in any moves towards a nuclear renaissance.

The European Pressurized Water Reactor (EPR) is a Franco-German collaboration that has developed a new pressurized water reactor system ready for immediate deployment.

The EPR project started in 1989 when the French company Framatome and the German firm Siemens created a joint company called Nuclear Power International (NPI). NPI was established to market the existing products of the two companies for export markets and more importantly to develop the EPR design so that it could pass a common regulatory approval process for both France and Germany [20, 21]. By combining the experience and resources of two large nuclear engineering firms the new NPI could claim involvement in more than 25% of global nuclear power generation, a figure higher than any competitor [20].

In 1999 following several years of successful collaboration between Framatome and Siemens in NPI and the design work on EPR, the two companies decided to fully merge their nuclear businesses to form Framatome ANP. Today, Framatome ANP is a joint activity of the Areva group and Siemens, providing a range of energy services, but with a particularly strong base in nuclear power systems.

The EPR design represents an evolutionary upgrade and modernization of both the French N4 reactor and the German Konvoi reactors [22]. The French units Chooz B1 and B2 (in the Ardennes) and Civeaux 1 and 2 are of the N4 design, while the German units Emsland, Isar 2 and Neckarwestheim 2 are of the Konvoi design.

Much of the financing of the EPR project is provided by the French utility Electricité de France and the German electricity utilities Preussenelektra, Badenwerk and RWE Energie. In the context of a possible UK nuclear renaissance it is interesting to note that between them EdF and RWE generated almost 20% of English and Welsh electricity in 2000/2001 (note Innogy is now an RWE company) [23].

The EPR is not characterized by the adoption of the types of passive safety systems favoured by BNFL/Westinghouse for the AP1000. Rather the EPR is an active reactor design building upon experience gained with the N4 and Konvoi models. Safety enhancements include a designed-in capacity to minimize adverse consequences arising from a core melt. Debris arising from a major and unmitigated loss of coolant accident would be prevented from reaching the bottom of the reactor pressure vessel by a core catcher [22]. The base of the reactor pressure vessel would be fitted with a melt plug (cf. later discussion of the ABWR) and in the event of a core melt this plug would give way and allow the molten core pieces to flow down a special channel into a prepared spreading compartment slightly displaced from the reactor. It is this compartment that is designed to catch the core. Such core catching concepts are not completely new, but they form a key part of the enhanced safety credentials of the EPR. The reactor also has enhanced containment compared with its precursors, and this gives far greater control over any gaseous fission product leaks from the core. The EPR design does incorporate one passive emergency feature, which is a capacity to flood the core in the event of serious over-heating.

From the start the proposed 1750 MWe units have been designed to use mixed oxide (MOX) fuels. The original proposal was for a 1500 MWe EPR fuelled with up to 100% MOX, but more recently designs have moved to a slightly higher power level operating with a maximum 50% MOX fuel contribution. If not using MOX, the uranium fuel would be low enriched at a level of approximately 5%.

As with many designs proposed for the first wave of a nuclear renaissance, economics has been a key concern of the EPR project managers. The original 1500 MWe design failed to match the projected lifetime costs of combined cycle gas turbine electricity generation and for this reason the designers were instructed to optimize the economic efficiency of the EPR proposal. This was the main driver in favour of a higher power (1750 MWe) design [24]. In addition the reactor has a design life of 60 years rather than the 40 years of previous European PWR systems. The EPR is also designed so as to require fewer refuelling interruptions and cycles of between 18 and 24 months are planned [25]. The EPR will use instrumentation and control based upon the N4 series of reactors.

Although established as a clear Franco-German project to be matched to the needs and regulatory concerns of those two countries, Framatome ANP is particularly conscious of the possibility of orders arising from any new nuclear build programme in other European countries. In December 2003 Framatome was able to announce a breakthrough that some believe heralds the start of a long-awaited European nuclear renaissance. The Finnish electricity company Teollisuuden Voima Oy (TVO) agreed to purchase an EPR plant for their Olkiluoto site in western Finland. The announcement of the Framatome ANP success followed the earlier Finnish parliamentary decision in May 2002 that paved the way for the building of a fifth Finnish power reactor, on economic, energy security, and environmental grounds [26]. See figure II.5.8.

Framatome ANP is following closely political developments in Eastern Europe. As several Eastern European countries continue on the path towards full European Union membership then so increases the pressure on these countries to replace their Soviet designed RBMK and VVER reactors with more modern designs.

In addition to Finland, Areva has recently started to stress business opportunities for the EPR in China and the United States [28]. Framatome ANP already has almost two decades of experience of doing business in China and has recently extended the range of technical cooperation with that country. The United States is a potentially large market in a nuclear renaissance and Areva have highlighted that fact. Despite that, however, Areva and Framatome ANP did not file for pre-licensing of their design with the US Department of Energy under the 2004 test licensing procedure for new nuclear power plants. Three consortia have proposed the AP1000, the ACR and the GE Advanced Boiling Water Reactor, but despite

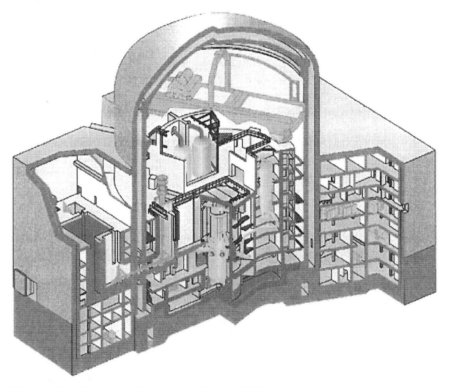

Figure II.5.7. Schematic of the Franco-German EPR nuclear power plant design. A reactor of this type has been ordered by the Finnish utility TVO (source: TVO).

Figure II.5.8. Artist's impression of the Olkiluoto site, Finland, when three nuclear power plants are operating. The plant in the foreground is the planned EPR plant that has recently been ordered by the Finnish electricity utility TVO (source: Areva [27]).

Table II.5.2. Technical characteristics of the European Pressurized
Water Reactor (EPR) (source: Framatome ANP [32]).

Thermal power	4250/4500 MW
Electrical power	In the 1600 MW range
Efficiency	36%
No. of primary loops	4
No. of fuel assemblies	241
Burnup	>60 GWd/t
Secondary pressure	78 bar
Seismic level	0.25 g
Service life	60 years

Framatome ANP's enthusiasm for the US market, no such pre-licensing has
so far been sought for the EPR in that country [29].

Any optimism in Framatome ANP from the Finnish success and potential in other markets must have been tempered somewhat by the political
trends in Germany. The last two German governments have been so-called
'red-green' coalitions between the Social Democratic Party and the Green
Party. The German Green Party has a long-standing hostility to nuclear
power generation, particularly new nuclear power plant construction. In
France there are more grounds for optimism about the possibility of new
nuclear build based upon the EPR. A recent French government study
showed that the EPR would represent a more economic source of electricity
generation than either coal or gas under a range of scenarios concerning
interest rates. The EPR option is the most cost effective even in the absence
of any carbon costs for the fossil-fuelled technologies [30]. In 2004 the French
utility Electricité de France (EdF) has been bringing forward plans to
develop a demonstration EPR in France. While the board of the company
has decided upon the strategy, no specific site for the new power plant has
yet been identified.

Table II.5.2 summarizes the main technical characteristics of the EPR
design. For a detailed assessment of how the EPR matches the needs of
European utilities the reader is recommended to consult the work of
Holger Teichel and Xavier Pouget-Abadie [31].

The EPR is not without its critics. On 15 December 2003 Greenpeace
activists occupied the offices of TVO in Helsinki objecting to the plans for
the construction of a new nuclear power plant in that country [33]. In
France the independent scientific commentators Global Chance have
produced a lengthy report. The English translation of the title of their
report is *The EPR Nuclear Reactor: A Useless and Dangerous Project* [34].
Thus far most of the criticism of the EPR plan seems to be an objection to
any form of new nuclear build rather than a criticism of this particular

technology. It will be interesting to see in due course how the economics of EPR with its particular technology choices compares with the economics of AP-1000 and ACR with a more distinctively modular approach emphasizing simplicity of design.

II.5.4 Boiling water reactors

The technology of boiling light water reactors (BWRs) is one that illustrates well the success of research, development and demonstration in the nuclear power field. From the earliest days of LWR research it had been clear that the separation of the reactor coolant circuit from the steam (and feedwater) circuit passing through the turbines is a source of inefficiency. In the case of a PWR both circuits are light water and the highly complex steam generators are simply there to transfer heat from the hot reactor coolant to produce steam to drive turbines. The reason that LWR designers did not immediately consider driving the turbines with light water steam directly from a coolant circuit was the worry that the light water in the coolant circuit should remain liquid at all times (as forced by the application of high pressure from the pressure vessel). Early LWR designers felt that it was vital to avoid steam voids in the reactor core (which would arise if boiling were to occur) as these would cause instabilities in the reactivity and that the reactor would be uncontrollable [35].

However, an important series of important BWR experiments were conducted at Argonne National Laboratory from 1953 onwards. These experiments, known as the BORAX experiments, demonstrated that under moderate pressure stable boiling resulted with no negative consequences for reactor control [36]. The BORAX-III test reactor went so far as to generate electricity that briefly in 1955 powered the town of Arco, Indiana, the first in the US to be powered by nuclear generated electricity. (The Soviet Union's Obninsk nuclear reactor has the best claim to be the world's first nuclear power plant as it generated its first commercial electricity on 27 June 1954. This water-cooled and graphite moderated reactor was finally shut down only in early 2002.)

The experimental boiling water reactor (EBWR) (figure II.5.9) was developed jointly by Argonne National Laboratory and General Electric (USA) at the Argonne site in Illinois. The EBWR operated from 1957 until 1967 running at 100 MWt and producing 5 MWe. The EBWR was used for much work on fuel development, including the possible use of plutonium in BWR fuel cycles.

As the decades passed the BWR concept continued to develop, so that by 1984 the LaSalle County Plant in Illinois entered service in the US, generating 1078 MWe. At the turn of the century more than 90 BWRs based upon the GE design are in operation in 11 countries supplying more than 4% of global

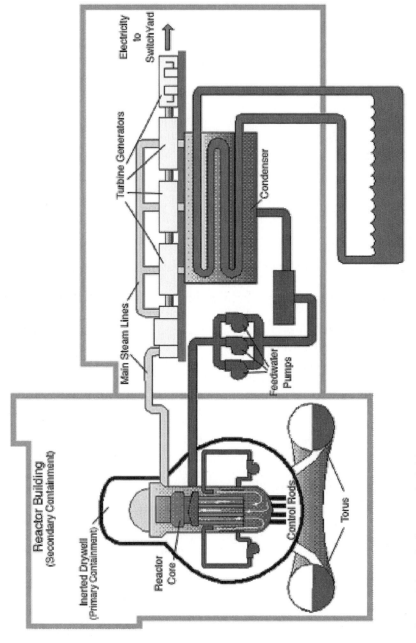

Figure II.5.9. Schematic diagram of a boiling water reactor (source: Tennessee Valley Authority).

electricity demand [37]. BWRs and PWRs dominate the load factor tables published regularly by the trade magazine *Nuclear Engineering International.* These LWR systems consistently provide very high levels of availability and reliability.

As noted above, the key benefit of the BWR approach is that the steam used to drive the electricity generating turbines is produced within the reactor pressure vessel directly from the water used in the reactor coolant circuit (see figure II.5.9). This eliminates the need for steam generators as used in PWR and CANDU reactors. Steam generators are very large and complex pieces of equipment and they represent a significant part of the engineering effort required to construct a PWR or CANDU system. As a PWR must ensure that its coolant is at all times in the liquid phase it is pressurized higher than is necessary in a BWR. This has the consequence that the BWR pressure vessel need not be as strong or as expensive to construct.

The use of a direct cycle is intrinsically more energy efficient. The heat capacity for a coolant determines the amount of heat that it can absorb for a given change in temperature. For the water in a PWR coolant circuit this is straightforward. For the coolant in a BWR there is a significant added benefit in that at the point of vaporization the heat capacity of the coolant increases discontinuously (the boiling of water is a strongly 'first order' phase transition). There is an increase in heat absorbed even for no change in temperature (this is the so-called 'latent heat'). This means that the steam emerging from a BWR is able to carry with it far more heat energy than the water leaving a PWR. For the same energy transfer the mass of water passing through the reactor core of a BWR can be significantly lower than in a PWR. The elimination of the need to transfer heat (with inevitable heat losses and inefficiency) from the coolant circuit to the separate circuit feeding the turbines is a further aspect of the increased efficiency of the BWR concept.

The intrinsic energy efficiency and simplicity of the BWR does not come without a downside. The main drawback of the BWR approach is that the water circulating in the reactor core must leave the containment building in order to feed the electricity generating turbines and to pass through the condenser. This water is inevitably somewhat radioactive and this means that substantial shielding is required for the steam and water pipes at all points including at the turbines and the condensers. Safety engineering is also more challenging as a loss of coolant accident not only has consequences for the stability of the reactor but also would cause concern regarding the path of the leaking water or steam. Ensuring worker safety has been a significant consideration throughout the evolution of the BWR concept.

The sources of radioactivity in the water of a BWR are twofold. The first, similar to the concerns surrounding radioactivity in the primary coolant circuit of a PWR, is the leak of minor actinides and fission products from any damaged rods of the nuclear fuel itself. In the absence of serious damage to the nuclear fuel the most likely vector for such contamination would be dissolved gaseous isotopes of xenon and krypton. The BWR, however, has the additional concern that any dissolved lubricants or corrosion products from the turbine (or from the condenser) may be activated by the neutron flux in the reactor core. BWR designers and operators are sensitive to this risk and great care is taken to ensure that the risk of such contamination is minimized. For instance, GE has made great efforts to ensure that cobalt has been eliminated from all parts of their modern BWR designs. Cobalt is a particular concern as it is easily activated to the highly radioactive isotope cobalt-60.

The design of a normative BWR is such that water at a pressure of approximately 70 bar is fed to the core, where about 10% is converted to steam within the core pressure vessel. The steam gathers naturally as a sepa-rated phase in a special region above the core [38]. The liquid water in the reactor core is mixed by means of a circulation pump, which extracts water from near the top of the water-filled part of the reactor core and returns it to the bottom. Two such circulation circuits are shown in figure II.5.9.

Hewitt and Collier report that the core power densities are roughly half of those in an equivalent PWR, although still significantly higher than those in gas-cooled reactors. They also note that the fuel elements consist of 3.5 m long bundles of zircalloy (zirconium alloy) canned low enriched uranium dioxide pellet fuel. Each fuel bundle is slotted into a vertical zircalloy channel of square cross-section in the core. As with a PWR refuelling requires that the reactor pressure vessel be partly drained and opened. As such, clearly refuel-ling can be done in conventional light water reactors only by taking the reactor off-line. The core design of the BWR makes refuelling somewhat more complex than in a PWR [39]. This is because refuelling in LWRs involves removal of the top section of the pressure vessel (the 'top head'). Changing the fuel bundles in their channels involves access to the core region below the steam separation region of the BWR pressure vessel. Partly to facilitate access to the core, the fuel bundles are smaller than the fuel elements of a PWR and therefore more fuel movements are required during a typical refuelling of a BWR.

The BWR concept has been noticeably successful in both operation and popularity with utilities worldwide. The popularity of the BWR grew throughout the latter part of the first wave of nuclear power plant construc-tion (mid 1950s to mid 1990s). As we look to the future we note that several BWR designs are ready for deployment early in a nuclear renaissance.

II.5.4.1   The advanced boiling water reactor (ABWR)

Of all the technologies poised for deployment in the first wave of a nuclear renaissance the ABWR stands out. All the design work and construction

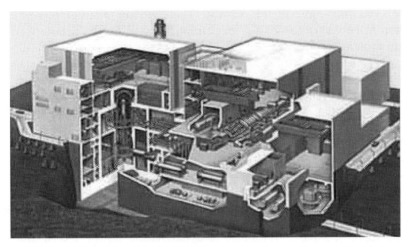

Figure II.5.10. GE–Toshiba–Hitachi ABWR cut-away diagram (source: Toshiba).

planning is complete and a fully operational and commercial example has actually been constructed. The ABWR is a GE design that has been implemented in Japan in collaboration with Toshiba and Hitachi (figure II.5.10). Two units were constructed for Tokyo Electric Power Company (Kashiwazaki-Kariwa units 6 and 7), coming on-line in the autumn of 1996 and the summer of 1997 respectively. Two further ABWR units are under construction in Taiwan. As with competing technologies, ABWR proponents are conscious of the need to minimize capital costs and economic risk in construction. Toshiba report that their capacity to construct ABWR systems is such that they can move from initial bedrock inspection to commercial ABWR operation in 39 months and this claim is backed up by actual experience at Kashiwazaki-Kariwa Unit 6 (1356 MWe) [40]. Toshiba claims that it was able to achieve the 39 months construction schedule by:

1. Expansion of parallel work, including building construction, equipment installation, electrical work and testing.
2. Reduced field work, including installation of larger pre-assembled components and expanded modularization.
3. Improved field productivity, including all-weather construction methods and extensive use of automatic welding machinery.
4. Total construction planning and management, including use of three-dimensional simulations and precise management control.

The circulation of light water moderator/coolant in the ABWR is achieved with ten internal recirculation pumps mounted on the 'bottom head'. While generally the BWR concept has been driven forward by GE in the US, the internal recirculation pump is a European innovation adopted by

GE for the first time in the ABWR. These reactor internal pumps have proven to be very reliable and they simplify the process of leak avoidance in the coolant-turbine circuit [41].

The ABWR construction plan involves heavy lift (1000 ton 'crawler' cranes [41]) so that pre-assembled modular elements may be inserted on-site. This approach mirrors that proposed by AECL for future construction of its Advanced CANDU Reactors. In addition to having improvements in construction, the ABWR has been designed for ease of maintenance, in particular of the reactor internal pumps and the automated fine motion control rod drives [41]. Removal of these devices from the core involves a special transport system and decontamination. This is facilitated by dedicated radiation-shielded service rooms where the necessary maintenance can be undertaken with minimal radiation doses to the workforce.

The ABWR design also confronts the particular need to prevent fluid leaks in BWR designs. While the pressure vessel can be less substantial than its PWR analogue, owing to the lower pressures involved, the containment of the BWR is far from insubstantial. First the reactor building and the turbine building are combined 'in-line' to form a very substantial edifice. The containment is a reinforced concrete containment vessel (RCCV) with a leak tight steel lining [41]. The outer shell of the reactor building is outside the reactor containment and acts as a second barrier. As is standard in reactor design, the reactor building has a slightly lower air pressure than outside in order to prevent any radioactive gases in the containment from escaping through small leaks to the outside world. The integration of the reactor and turbine buildings is an extremely cost effective way to improve the seismic resilience of the plant. Japan is noteworthy for its enthusiasm for civil nuclear research and development, but also for its unstable geology. Seismic resilience is a crucial concern in Japan and several other markets considering a nuclear renaissance.

The ABWR design incorporates three completely and physically separate cooling systems. The 'reactor core isolation cooling system' is even powered by reactor steam so that it can operate in the absence of all of the multiple sources of electrical power designed into the plant. Among the various requirements of safe operation, one stands out—the need to keep the core covered in all circumstances. As with all modern LWRs much thought has been given to this. The ABWR designers tasked themselves with matching the long operator intervention times associated with modern planning for loss of coolant accidents (LOCAs). The ABWR achieves this through the use of automated active plant responses, which have been designed to ensure that no operator intervention will be required in the first 72 hours of a LOCA [41]. Plant operators can be reassured that no immediate action is required in the event of a LOCA. In most cases there would be sufficient time to gather and fly in a national team specially trained and held in readiness to manage such eventualities.

The ABWR designers have truly considered worst-case situations. The risk of 'melt down', to use the now ubiquitous popular term, or a 'core on the floor' accident as the ABWR designers term it, would represent perhaps the worst situation that might occur with an LWR (cf. discussion of the European Pressurized Reactor). In such a scenario the reactor core becomes uncovered and then melts to the floor of the reactor pressure vessel in an uncontrolled way, settling in an unknown state. One thing can be certain—if the reactor is in trouble in this way, then the core debris would be hot. The ABWR designers have incorporated a fusible valve, which would be opened by the heat of the core debris. This would release water from the suppression pool to quench the core debris and would also greatly reduce the amount of gas chemically produced if the core material were ever to come into contact with concrete [41].

The ABWR also incorporates a pressure burst disk to release excess gas from the pressure vessel in the event of any dangerous over pressurization. Any gases liberated this way would not simply vent into the containment, but would be routed into the suppression pool. Only if the airspace above the suppression pool became over-pressured would a further venting occur. Even in that eventuality the two step emergency venting would capture a high fraction of the emitted radioactivity [41].

The control room of the ABWR adopts a best practice regarding safety critical instrumentation (such as used in air traffic control). For instance, the entire system could, if need be, be operated from a single console. To facilitate this any operator can call up any aspect of the system to their own console.

The long history of development of the ABWR (the first discussions were held between GE and possible international partners as far back as the summer of 1978) has not been without its setbacks. The Toshiba and Hitachi joint plans to construct a pair of ABWR units at Ashishama ended with the cancellation of the project. The cancellation followed widespread public opposition, but more importantly it came after the withdrawal of support by the head of the Mie prefecture [42].

II.5.4.2 The Framatome ANP SWR1000

In terms of reactor systems research the contribution of Siemens Nuclear Division to the merged Framatome ANP is noteworthy for including a design for a passively safe BWR known as the SWR1000 [43]. This is a collaborative design project undertaken by the Framatome ANP GmbH in partnership with German electric utilities and European partners. The philosophy is to control accidental reactor transients through the use of systems based entirely on the laws of nature (gravity etc.) and as such this may be compared with the philosophy of the BNFL/Westinghouse AP1000 discussed earlier. Active systems will, as far as possible, employ technologies that have already been demonstrated to be safe and effective in previous reactor designs. Once again, the overriding challenge to the design team has been to reduce the capital cost of nuclear power while preserving or enhancing reactor safety.

II.5.4.3   Simplified boiling water reactors (SBWR)

In addition to the central role General Electric in the ABWR (see above), it has also devoted much work to other passively safe boiling water reactor concepts including, most notably, the 600 MWe simplified boiling water reactor (SBWR). Key to the SBWR concept was that it would use natural liquid water circulation in the core region. As such, no recirculation pumps would have been required. Also the decay heat removal systems were designed to rely on natural convection cooling alone rather than on active systems. However, the SBWR as originally conceived is now most unlikely to form a part of any nuclear renaissance as the project was cancelled by GE in 1996 [44], as its interest returned to larger reactor types, such as the 1390 MWe ESBWR.

One variant of the small-scale SBWR concept, however, survives in the form of the GE and Purdue University Modular Simplified Boiling Water Reactor (MSBWR) project. This is a project closely associated with the US led Generation IV activity (see chapter 8). The aim is to produce two small reactor designs, one at 200 MWe and an even smaller one at 50 MWe. Like the SBWR, on which it is based, the MSBWR will employ passive features such as convection cooling. It will use 5% enriched BWR fuel with a ten-year refuelling interval. The MSBWR is designed for minimal intervention and maintenance [45]. This facilitates deployment of the reactor in remote areas far from existing electricity transmission grids or in clusters with several other similar units where a single skilled labour force can support the operation of several clustered units.

The United States was not the only country to pioneer the early development of boiling water reactors. The Soviet Union was also acutely aware of the advantages of such an approach, but the Minatom experiences were shaped by the fact that they did not have early access to the results of the US Borax experiments. The consequence was that the unique Russian RBMK design is a boiling water reactor with a graphite moderator. The use of a graphite moderator was intended to circumvent the two-phase difficulty that the Americans originally expected with a boiling moderator. In fact the use of a graphite moderator and a boiling light water-coolant gave rise to a significant safety problem that was eventually to be the undoing of the design. The RBMK reactor design suffers from a positive void coefficient when steam voids form within the reactor core. It is this aspect that lies at the heart of the disaster at the Chernobyl Unit 4 plant in 1986. G F Hewitt and J G Collier provide a thorough account of the Chernobyl accident in their book *Introduction to Nuclear Power* [46]. The RBMK is an extremely large and capital-intensive design, but it was the accident at Chernobyl that ensured that the Russian graphite moderated boiling water reactor technology will not figure in a global nuclear renaissance. That does not mean, however, that Russia has nothing technological to contribute to a renaissance. In the area of fast reactors, for instance, Russia has much useful knowledge and experience and we shall consider this further in chapter 7.

Chapter 6

High-temperature gas-cooled reactors

Critical fission reactor technologies of 2010 and beyond are likely to include high-temperature gas-cooled reactor (HTGR) systems. As with other renaissance technologies, HTGR technology has roots stretching back several decades.

Two experimental reactors in particular are worthy of mention. The first was the Dragon Reactor Experiment at Winfrith, UK, which operated between 1964 and 1975. Funded and managed as an OECD/NEA international joint undertaking, the primary function of the reactor was for materials testing of both components and fuels [47]. Much of the current data informing high-temperature reactor design comes from experience gained on the Dragon machine. In particular the Dragon experiment should be remembered for its pioneering use of helium as a reactor coolant and for its use of coated fuel particles (of highly enriched uranium–thorium carbide) [48]. The Winfrith developments were accompanied, and followed, by related endeavours at Jülich in Germany where a prototype HTGR known as the Arbeitsgemeinshaft Versuchsreaktor (AVR) (figure II.6.1) was operated from 1967 until 1988 [48]. The AVR pioneered the 'pebble bed' concept to be discussed later, but faced difficulties with fuel quality and fuel handling. One key lesson from the AVR was learned the hard way. It became clear that future pebble bed designs should avoid the need for control rods to move amongst the fuel pebbles, as damage to the fuel pebbles could easily arise in such designs. Modern pebble bed reactor designs have the control rods operating within the reflector rather than within the fuel assembly.

These experimental studies in the UK and Germany were followed by attempts to produce commercial power plants based upon high-temperature reactor technology. One was the German 300 MW designed thorium high-temperature reactor (THTR-300) pebble bed facility. The THTR-300 suffered from licensing and funding problems which forced its early closure [48].

Figure II.6.1. The AVR in operation at Jülich, Germany (source: www.pbmr.co.za).

The experience associated with the attempted commercialization of HTGR technology in the United States provides one of the most informative and illustrative stories of the difficulties facing the first wave of nuclear power development. The problematic project in question was the Gulf General Atomics commercial HTGR at Fort St Vrain, Colorado.

The Fort St Vrain project was beset with engineering difficulties from the start of construction in 1968 until the closure of the plant in 1989. Many but by no means all of the difficulties concerned the leakage of water from the bearings of the helium circulators into the helium gas coolant circuit. This was made all the more problematic: the design did not incorporate a mechanism by which the core could be dried out expeditiously. For a history of the Fort St Vrain reactor the reader is recommended to consult the website developed by past and present workers at the plant [49].

In some ways the Fort St Vrain saga is a negative allegory of the entire history of nuclear power in the United States. Following its failure as a nuclear power plant, the Fort St Vrain reactor system was decommissioned and removed, to be replaced by a set of three natural gas-fuelled gas turbines together producing 720 MWe at high levels of reliability and low cost. The Fort St Vrain experience encompasses the early enthusiasm for innovative reactor systems, the increasing disenchantment with nuclear power through the 1970s and 1980s, and the move to natural gas in the 1990s.

The early HTGR experience exhibits the symptoms of technological over-reach characteristic of the first wave of nuclear power development. Similar examples of excessive ambition and engineering hubris are encountered elsewhere in this book (e.g. the fast reactor experience). The ambition and vision of the General Atomics engineers at Fort St Vrain matched the then state of the art in terms of nuclear fuel engineering and core design, but crucially went beyond the state of the art in terms of high-temperature gas handling. Similarly the AVR experience in Germany illustrated state of the art nuclear engineering, but required materials science capability beyond that available at the time.

The early HTGR developments were made possible by the enthusiasm of the first wave of nuclear power plant construction, but the science and engineering were not yet ready for such technologies. In the past 30 years fluid flow engineering and materials science have both improved significantly so that we can now say with confidence that the early promise of the HTGR concepts should now be realizable.

Recent engineering developments are likely to make possible technologies that seemed excessively problematic when HTGRs were first attempted. The most important development has probably been the appearance on the market of reliable high-temperature gas turbines which are able to withstand the rigours of operation in the hot helium gas stream emerging from the reactor.

All variants of the HTGR concept are characterized by the use of enriched uranium fuel (although exotic fuel cycles may also be used), graphite moderation, and helium cooling. Typically the graphite for the neutron moderation of the core forms an integral part of the fuel assembly.

Future HTGRs will build upon a wealth of successful experience in the development and operation of gas-cooled reactors. The UK in particular developed a series of Advanced Gas-cooled Reactors, the AGRs, in the 1970s and 1980s. These large graphite moderated and carbon dioxide cooled reactors initially faced construction delays and cost over-runs, but they have for the most part performed well (recent experience excepted) since the privatization of the UK electricity industry in the early 1990s. However, HTGRs are both more technically challenging and more cost efficient than the more conventional gas-cooled reactors such as the British AGR and its gas-cooled predecessor the Magnox design of the 1960s.

At present there are two leading variations on the HTGR theme—the 'pebble bed' and the 'prismatic' reactor. Although both variants share many common attributes, we shall be considering them separately. Both variants have been developed: a direct cycle (the helium coolant drives the electricity generating turbine directly) and an indirect cycle (in which the heat from the helium is exchanged with water to raise steam to drive the turbine). Wang *et al* discuss the relative advantages and disadvantages of the two approaches and include the point that, despite maximal separation of the turbines and the reactor in the direct cycle schema, there is still a greater risk in direct cycle designs from the disintegration of the turbine or the compressors when rotating at very high speed—so-called 'missile damage' [50].

One key common attribute of the majority of HTGR proposals is the fuel design known as Triso after the three (TRI) separate layers of graphite isotropically (ISO) cladding each fuel particle. This fuel/moderator combination has undergone substantial development and it is important to consider it in some detail as this technology differs significantly from the fuel/moderator thinking involved in the first generation of commercial nuclear power plants.

II.6.1 Triso fuel

At the heart of HTGR fuel systems are many millions of sub-millimetre fuel kernels known as Triso fuel (figure II.6.2). Within the Triso fuel particle and surrounding the nuclear fuel itself is a shell of porous carbon able to absorb gaseous fission products and a layer of pyrolytic carbon specially developed for the nuclear industry for high thermal stability and fire resistance. Outside these carbon layers each Triso particle has a silicon carbide barrier coating, which provides the strength and the protection against fission product and minor actinide release. One consequence, however, of this silicon carbide layer is that the fuel pebble is so resistant to chemical attack that a once-through fuel cycle seems inevitable for HTGR systems employing Triso fuel.

One drawback of an integrated fuel and moderator assembly used in Triso fuels is that the volumes of radioactive waste spent fuel produced are greater than from either conventional reprocessing based fuel cycles or from a once-through light water reactor fuel spent fuel waste stream, although it is thought that HTGR wastes may be more densely packed in a repository than is possible for spent LWR fuel. These issues are considered later in this chapter in the context of the PBMR project. One of the key arguments of those proposing a nuclear renaissance has been that future nuclear wastes arising from renaissance technologies will be a negligible fraction of the total future waste inventory. If HTGR approaches based upon Triso fuel produce larger volumes of waste then, in this respect at least, they will be less attractive than their light water reactor analogues.

It is perhaps conceivable to imagine deliberately crushing Triso fuel in order to gain access to the nuclear fuel itself for reprocessing, but this would seem to eliminate a key benefit of Triso-based systems—that is, all being well, radiologically harmful materials remain encased within the

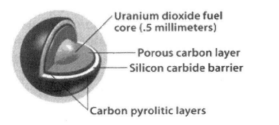

Uranium dioxide fuel core (.5 millimeters)

Porous carbon layer

Silicon carbide barrier

Carbon pyrolitic layers

Figure II.6.2. Construction of a Triso fuel kernel (source: *Technology Review* [51]).

highly impermeable silicon carbide shell. Surrounding the SiC shell is a further layer of pyrolytic graphite. It is the three separate layers of graphite which provide the necessary moderation of the fast neutrons emitted by the fission of the enriched uranium in the Triso fuel.

II.6.2　Pebble bed HTGR

The pebble bed concept relies upon the elegance and simplicity of the idea that the fuel is self-organizing in the core and that refuelling and de-fuelling are continuous and on-power.

The fuel for pebble bed reactors consists of spherical fuel assemblies a little larger than a billiard ball. These are the 'pebbles' of the pebble bed concept. Each pebble contains around 15 000 Triso fuel particles, so that the fuel zone at the heart of each spherical fuel assembly, or 'pebble', contains about 9 g of low-enriched uranium [52] (figure II.6.3).

The real elegance of the pebble bed concept arises from the organization of the fuel pebbles in the reactor core. Traditionally nuclear reactor core designers have devoted much care and effort to the exact positioning of fuel rods in precisely engineered channels. Significant outages have occurred when fuel rods have become stuck or broken during refuelling operations in such reactor designs. The elegance of the pebble bed approach is that there is no deliberate or active arrangement of the fuel pebbles at all. They are simply

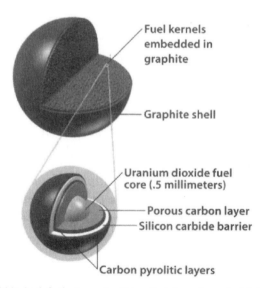

Fuel kernels embedded in graphite

Graphite shell

Uranium dioxide fuel core (.5 millimeters)

Porous carbon layer

Silicon carbide barrier

Carbon pyrolitic layers

Figure II.6.3. Pebble bed fuel element—Triso fuel kernels embedded in fuel 'pebbles' (source: *Technology Review* [51]).

Fig II.6.4. Gumballs in a traditional vending machine.

poured into the reactor core and allowed to arrange themselves in random close-packed arrangement like gumballs in an old-fashioned American gumball machine (figure II.6.4). The beauty of this arrangement is that it is precisely known how much graphite there will be between two spheres of pebble bed fuel.

Any arrangement of the pebbles will inevitably leave ample space for the helium gas to pass around the pebbles and to cool them. The excellent thermal conductivity of the pyrolytic graphite parallel to the surface of the fuel pebble should help prevent problematic hot spots from developing in the core.

The design relies upon the integrity of the fuel pebbles, with few fracturing or disintegrating in the core. The cooling and moderation of the reactor require this but, importantly, fuel kernels emerging from broken fuel pebbles are unlikely to have their fast neutrons sufficiently moderated to favour fission in neighbouring pebbles.

The Triso fuel encased in the pebbles enters the reactor system via the fresh-fuel tank shown to the lower right of figure II.6.5. These new fuel elements pass the fuel burn-up assay stage (the element immediately above the spent-fuel container in the figure) before being routed into the top of the reactor vessel. There the fuel rests on the top of the self-organizing stack of fuel pebbles. The pebble bed is topped-up continuously with the reactor at full power. The on-load refuelling of the reactor allows for greater reactor efficiency and reliability. While new pebbles are added to the top of the reactor, used pebbles are removed from the bottom. These pebbles are inspected for damage and pebbles with signs of damage are diverted to a special broken pebble container. The overwhelming majority of the pebbles

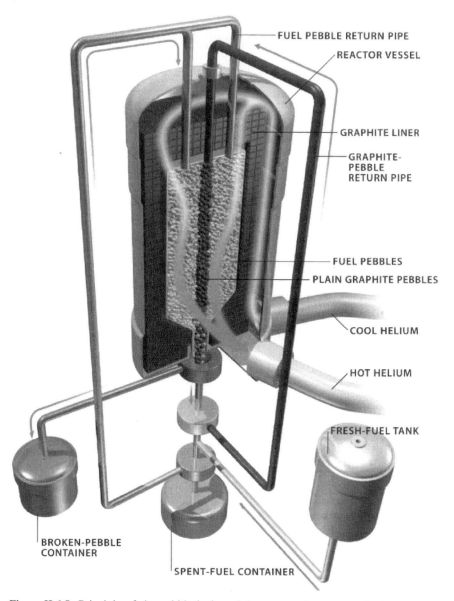

FUEL PEBBLE RETURN PIPE

REACTOR VESSEL

GRAPHITE LINER

GRAPHITE-
PEBBLE
RETURN PIPE

FUEL PEBBLES

PLAIN GRAPHITE PEBBLES

COOL HELIUM

HOT HELIUM

FRESH-FUEL TANK

BROKEN-PEBBLE
CONTAINER

SPENT-FUEL CONTAINER

Figure II.6.5. Principle of the pebble bed modular reactor (source: *Technology Review* [51]).

must be undamaged for the reactor design to fulfil its potential. Undamaged pebbles are then distinguished from the plain graphite pebbles used to ensure proper and sufficient moderation in this version of the design. The plain graphite pebbles are directly routed to the top of the reactor and routed down a central channel within the reactor. The most important benefit of the pebble bed fuelling and de-fuelling system is that the reactor is kept at full power throughout the process. This minimizes downtimes for the reactor and maximizes availability of electricity to the local community. The economic benefits to the utility company are clear. More important, however, is the continuous availability of electricity, as one of the key anticipated markets for small modular pebble bed systems is for electricity supply for small communities far from existing electricity infrastructures. For these communities any periods of time without the pebble bed reactor could constitute periods without any electricity at all.

The pebble bed core is surrounded by a graphite-lined reactor vessel. The graphite liner acts as a neutron reflector. In modern pebble bed designs criticality is controlled by adjusting the neutron reflectivity of the reactor lining. By inserting control rods into the graphite liner (not shown in figure II.6.5) rather than into the fuel itself, damage to the fuel pebbles is avoided. Specific pebble bed projects are discussed in the following sections.

II.6.2.1 The pebble bed modular reactor, South Africa

The pebble bed modular reactor (PBMR) demonstration project centred at Koeberg 30 km north of Cape Town in South Africa is the most talked about HTGR project positioned to contribute to the nuclear renaissance. It is neither the most far advanced HTGR project (both the HTR-10 and the HTTR discussed later are more developed) nor is it the most technologically ambitious (that honour might best be held by the GT-MHR). However, the PBMR is the HTGR project with the greatest visibility and commercial momentum behind it.

The PBMR is a consortium project headed by the South African utility company Eskom and involves BNFL of the UK and the Exelon Corporation of Chicago, Illinois. A second South African partner (IDC, the South African Industrial Development Corporation) is also involved.

Perhaps the greatest benefit of the South African PBMR variant of the pebble bed concept is that at no point in the electricity generation process is steam raised to drive the turbines. Rather the helium gas used to cool the reactor is directed directly through an electricity-generating turbine capable of producing 110 MWe. The helium emerging from the pebble bed is, as the HTGR name suggests, at a high-temperature. Using high-temperature gas emerging from the reactor core directly to drive the electricity generating turbines is in some ways reminiscent of the boiling water reactor innovation in LWR systems. The associated challenges in gas-cooled reactors, however, while simplified by the inert nature of the gas, are made much more demanding by the higher temperatures involved.

One unusual aspect of the PBMR design is that unlike orthodox nuclear power plant operations it is not designed to be a base load power source.

That is, it is not expected that the PBMR will operate at constant full power during its availability. Rather, the PBMR is expected to be 'load following' in that it will adapt its output to match fluctuating demands. The ability to follow a fluctuating demand is very important in permitting the PBMR to serve its primary objective of bringing high quality power to isolated communities far from electricity transmission infrastructures. Without a large-scale electricity grid in which one can aggregate fluctuating supply and demand, the ability for supply and demand to be matched by load following becomes extremely important if economic and efficient power generation is to be achieved. The alternative to load following would be to develop large-scale cost-effective energy storage technologies. This is a difficult and expensive proposition of importance not only to nuclear power plants in isolated environments but, more importantly, to renewables. There is currently much work under way internationally into energy storage, but at present the emphasis of the PBMR project on a load-following capability seems to be well thought out.

The direct drive cycle of the helium cooling gas is described as a 'Brayton cycle'. The Brayton cycle is a standard idealized engineering thermodynamics process for the extraction of energy from hot high-pressure gases. As such, the Brayton cycle is a member of the family including the Carnot cycle, the Otto cycle, the Diesel cycle and the Stirling cycle of introductory engineering thermodynamics.

The Brayton cycle is a four-leg cycle. As it is a closed circuit we might consider starting our description of the path of the helium at any point but, for sake of argument, let us start the cycle with the initial pressurization and pre-heating of the helium gas, as shown in the lower half of figure II.6.6. The first leg consists of an adiabatic (i.e. no flow of heat to or from the gas) pressurization and compression of the helium gas followed by pre-heating in the recuperator. The gas then enters the top of the reactor at 500 °C where it is heated in the core and allowed to expand downwards at a constant pressure of 9.0 MPa until it reaches its highest temperature of 900 °C (second leg). The gas is then adiabatically released and allowed to expand and depressurize through the sequence of gas turbines, the first two of which drive the compressors used for the first leg pressurization and the third of which is the power turbine which generates the electricity. This adiabatic depressurization is the third leg of the cycle. The fourth leg is a constant pressure decrease in volume as the remaining heat is released from the gas. The fourth leg of the cycle is achieved by passing the gas through the primary stage of the recuperator, the secondary side of which facilitated the pre-heating in the later part of the first leg of the cycle. Even after passing through the primary of the recuperator, the stage 4 helium is still too hot to re-enter the cycle. This helium is therefore cooled further in an 'intercooler'. Once cooled back to 500 °C, the helium is at a sufficient density for it to be well suited to efficient compression in the first leg of the process,

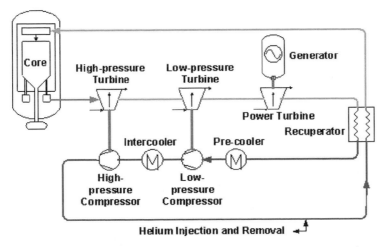

Figure II.6.6. The Brayton cycle for the South African PBMR (source: www.pbmr.co.za).

and so the process repeats itself. Clearly the process does not consist of a single pulse of gas circuiting the system, but rather all parts of the cycle are served continuously and are happening simultaneously.

It is only in recent years that improvements to bearings have been achieved to the extent that reliable gas turbines are now available that can operate in a 900 °C high-pressure helium flow. These aspects are discussed further in the context of the GT-MHR prismatic HTGR. The increased efficiency, and the greater simplicity achieved by eliminating steam from nuclear power, are key advantages of Brayton cycle based HTGR concepts.

The basic PBMR Brayton cycle described above may be related to the structure of the PBMR system by means of the schematic in figure II.6.7. The four main interconnected pressure vessels consist of the reactor vessel, the two turbo-compressors that make possible the first and third legs of the Brayton cycle, as described above, and the fourth pressure vessel holds the power turbine and electricity generator.

The schematic (figure II.6.7) fails, however, to capture the actual look of the PBMR system and this is shown in figure II.6.8. The design shown in that figure illustrates well the combined engineering challenges of the Brayton cycle helium gas system and the pebble bed fuel management system.

Recent modifications to the PBMR design include a redesign of the central reflector, which originally had been envisaged as a dynamic column of graphite spheres (see figure II.6.5). The revised design includes a fixed graphite column as the central reflector. This change has simplified reactor operations and allowed for a reduction in the peak operating fuel temperature to below 1130 °C.

A detailed feasibility study for the South African PBMR was completed at the end of 2002 and during 2003 a review of the business case was under

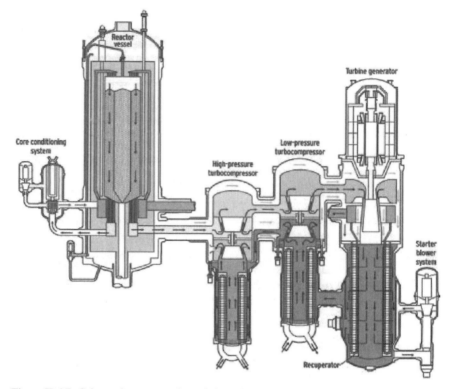

Figure II.6.7. Schematic cross-section of the Eskom PBMR (source: www.pbmr.com).

way. In addition, the Environmental Impact Assessment process was completed in October 2002 and the files were passed to the South African Department of Environmental Affairs and Tourism (DEAT) for consideration. In June 2003 DEAT ruled in favour of the PBMR and its location at Koeberg near Cape Town. In response several environmental organizations have started to challenge the PBMR programme in court.

II.6.2.2 MIT's modular pebble bed reactor

The MIT modular pebble bed reactor (MPBR) is an indirect cycle, helium-cooled reactor system designed to adhere to, and to surpass, US regulatory norms. The MPBR reactor core is similar in design to that of the South African PBMR project and there has been much sharing of insight and experience between the two teams. That said, the industrial consortium under-pinning the PBMR contrasts with the more academic (university and national laboratory) culture of the MPBR project.

The MPBR has undergone a series of design revisions, but the following attributes are fundamental to the project. The MPBR is a pebble bed reactor

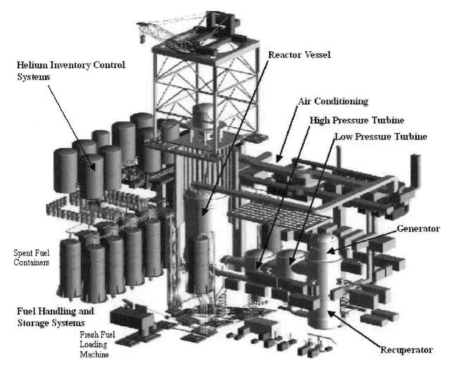

Figure II.6.8. Overview of the Eskom PBMR design (source: www.pbmr.com).

with a high degree of modularity and factory-based construction. The design philosophy has been driven by:

- a desire to minimize capital and operating costs,
- an intention to avoid potentially show-stopping engineering challenges and
- a desire to achieve an inherently safe design.

With regard to low cost construction, operation and maintenance, the MPBR is designed to require no more land than an equivalent conventional (e.g. natural gas fuelled) power plant [52]. The construction of the MPBR relies on a highly modular approach. The modules must be small enough (8 × 12 × 60 ft, ~2.5 × 36.5 × 18 m, maximum) to be transported on the back of a US 200 000 lb trailer truck.

In a key departure from the philosophy of the South African PBMR team, the MPBR design separates the reactor high-pressure cooling circuit from the electricity generating turbine circuit [50]. The primary reason for this separation is a concern that insufficiently high quality turbo-machinery is currently achievable for a single cycle design. The beneficial consequence of a move to an indirect cycle is that the rotational speed of the MPBR shafts is significantly lower, at several hundred revolutions per minute, than in alternative designs such as the PMBR or the GT-MHR (discussed later) where speeds of several tens of thousands of rpm are required for efficient operation [50].

Somewhat unusually the MIT MPBR design has horizontal shafts for the compressors and the electricity-generating turbine. The horizontal geometry facilitates the use of magnetic bearings and also adds to ease of maintenance. Originally the MIT team proposed a two shaft turbo-machinery design for the MPBR (one for the electricity generating turbine and one for the compressors) [52]. However, by increasing the number of separate shafts in the compressor set it becomes easier to construct the turbo-machinery not from next generation technology but from close to off-the-shelf technology. In this context three-shaft [52] and four-shaft [50] designs have been evaluated.

The MPBR design team aim to combine the reliability of industrial strength turbo-machinery with the efficiency of aircraft gas turbines [52]. With a multiple shaft turbo-machinery concept and the use of an indirect cycle these ambitions would seem to be easily achievable. The technical drawback in deciding to move to an indirect cycle is the need to employ an intermediate heat exchanger between the two circuits and this component has the potential to become the greatest technological challenge in the design of the reactor system.

In the MPBR design both the primary reactor coolant circuit and the secondary electricity-generating circuit (operating on a Brayton cycle) are helium gas circuits. The use of helium rather than steam in the secondary circuit permits higher efficiencies, as the secondary circuit can operate at far higher inlet temperatures than would be possible with steam. The use of helium in the secondary circuit also avoids well-known problems of corrosion associated with high-pressure and high-temperature steam cycles [50].

Rated at 110 MWe, the MPBR is characteristic of the HTGR move away from the dominant nuclear power paradigm of large base load centralized generation. For an extensive overview from December 2001 of the MPBR and related concepts, the reader is recommended to consult a report from the Idaho National Engineering and Environmental Laboratory edited by William K Terry [53].

II.6.2.3   China's HTR-10 high-temperature reactor

China has adopted an HTGR research programme to complement its main power reactor programme based around PWR technology. In addition to the HTGR technology attributes for high efficiency electricity generation, China's policy makers have been persuaded of the importance of HTGR technology for the following reasons [54]:

- to permit nuclear power generation in those regions lacking suitable locations for PWR construction,
- to provide process heat for heavy oil recovery and the petrochemical industry and
- to provide a process heat resource for coal gasification and liquefaction.

The emphasis on process heat is similar to the Japanese motivation behind the HTTR project to be discussed later in this chapter. What is particularly interesting and unusual, however, is China's interest in applying HTGR nuclear heat to improve the efficiency and cleanliness of fossil fuel energy sources, which are abundant in China.

China intends to follow research at its HTR-10 facility with a series of modular HTGRs to complement its PWR nuclear programme. While nuclear power has continued to develop and grow in China over the last two decades of the 20th century it is important to note that China intends to expand its use of nuclear power during the period of the western 'renaissance' discussed in this book.

The HTR-10 reactor project has been led by the Tsinghua University Institute of Nuclear Energy Technology (INET). The reactor has been constructed at Changping, about 40 km north-west of Beijing city centre. The project was approved by Chinese regulators in December 1992 and achieved first criticality eight years later in December 2000.

The HTR-10 is a design test bed for modular HTGRs of the pebble-bed type [55, 56]. The reactor core and the steam generators are housed in two separate pressure vessels, which are arranged side by side and interconnected by a hot gas duct carrying helium at approximately 700 °C. The two pressure vessels together constitute the primary pressure boundary of the reactor system. The two pressure vessels are designed to operate hot, as much of their metal surface is in contact with 3.0 MPa helium at its cooler return temperature of approximately 250 °C.

The reactor core has a diameter of 1.8 m and a mean height of 1.97 m. It is surrounded by graphite reflectors. The HTR-10 has a thermal power of 10 MW. The outer reflector blocks are boronated to provide thermal and neutron shielding to metallic internal components and to the reactor pressure vessel [55].

The pebble bed spherical fuel elements are fabricated from Triso fuel. Each HTR-10 fuel pebble contains about 8300 Triso particles totalling in all 5 g of 17% enriched uranium. The reactor core contains approximately 27 000 fuel elements, or pebbles, which circulate in multiple passes through the core. The pebbles emerging from the reactor are moved using a pulse pneumatic fuel handling system. Each fuel element is assessed individually for damage and nuclear burn-up. Those that have not yet reached the desired burn-up level are sent back pneumatically to the top of the reactor core [55].

Visions of the renaissance II: Powering isolated communities 2020

By 2020 a new generation of small modular high-temperature gas-cooled reactors is supplying electricity in geographically isolated areas, most notably in Russia, South Africa and China. These countries combine high technology energy intensive ambitions with under-developed electricity distribution infrastructures. Rather than devote huge resources to bringing transmission grids to all corners of their vast countries these governments have spent the first two decades of the 21st century developing decentralized electricity generation capacity, including nuclear capacity. Within the nuclear sector these countries have found small modular HTGR systems particularly beneficial. Unsurprisingly in South Africa the technology adopted has been Eskom's pebble bed modular reactor (the PBMR) while in Russia similar efficiencies have been achieved with the gas turbine–modular helium reactor (the GT-MHR).

High-temperature gas-cooled reactors have not made the same impact in Europe and North America. This has largely been due to articulate opposition from anti-nuclear campaign groups and limited resources and capacity within key regulatory bodies. The regulatory and political hurdles and resource difficulties have led to delays and only in the period 2010 to 2020 have such bodies been in a position seriously to address the possibility of new HTGR build. Some new build proposals have, however, been approved in this period and it is expected that, while full scale power generation is not occurring in these regions by 2020, there is optimism that such power will be flowing by 2025.

It has been interesting to see how western nuclear companies have played a leading role in helping HTGR power generation in Russia, China and South Africa while the same companies have achieved far less success in their home countries. It is thought by many that the HTGR story is one of developing country innovation dominating over western technological inertia. However, the fact that western companies are intimately involved in the innovation process rather complicates this view.

The HTR-10 has two independent reactor shutdown systems [55]. The first consists of ten control rods in the side reflector. As in other modern pebble bed designs there is no design requirement for reactor control mechanisms within the pebble bed core. The second system also acts in the side reflector and consists of seven slotted holes of $160\,mm \times 60\,mm$ that may be filled rapidly with small balls of neutron-absorbing material.

Importantly the HTR-10 can be brought to a cold shutdown solely by one of these two systems acting in isolation. Once the shutdown system has ended the nuclear reaction in the core there is no need even for the circulation of helium coolant. The HTR-10 has strong negative temperature coefficients and decay heat can be safely removed without the need for circulating helium coolant. The reactor systems are designed to cope straightforwardly with a loss of coolant flow during full power operations.

The HTR-10 relies upon an intermediate heat exchanger and an integrated steam generator for heat removal. The HTR-10 itself is capable of generating 2.5 MWe from turbines driven by steam at 400 °C and at 4.0 MPa. In addition to electricity generation, the water/steam secondary circuit is optimized to produce hot water for district heating use [55].

As with the Eskom PBMR system discussed earlier, the HTR-10 employs no standard reactor containment. Beyond the double pressure vessel the only barrier to the release of radioactive materials in the event of a serious accident is the concrete reactor building, but this is neither a pressure retainer nor a leak-tight containment in the true sense. As shall be discussed in the next section, the lack of reactor containments in many HTGR designs is regarded by many as a key weakness.

II.6.2.4   Criticisms of the pebble bed concept

The pebble bed concept, despite having many enthusiasts, is not without its critics. One of the most eloquent and knowledgeable is Edwin S Lyman of the Nuclear Control Institute in Washington, DC [57]. In 2001 he characterized the drawbacks of the pebble bed modular reactor variant, and the comments below are informed by Lyman's article.

- The PBMR lacks a robust containment and instead has a filtered vented building. This reduces costs, favours modularity and allows for the addition of further units, and facilitates natural convection cooling in emergency situations. The lack of a proper containment is regarded by some as a potentially dangerous lack of an important safety barrier to the release of radioactive materials in the event of a serious accident or terrorist attack.

- HTGR systems make use of fuel/moderator composites usually in the form of Triso fuel. In pebble bed reactors the Triso fuel is assembled in spherical fuel elements of a few centimetres diameter coated in pyrolytic graphite. The concern of the critics relates to the possibility of a graphite fire. The PBMR is designed to operate at approximately 900 °C with the graphite-clad fuel elements in an inert helium atmosphere. The concerns become substantial if air or water were to encounter a bare graphite surface at such temperatures. It is important to note that a fire of the outer graphite cladding of the fuel pebbles probably would not in itself lead to a breakdown of the SiC-clad Triso fuel or the release of radio-active materials. The designers are confident that the ingress of oxidizers (air or water) would always be inhibited by the nature of their design and that the barriers to release could cope with the gaseous and sooty products of large scale graphite combustion. Experience at the Jülich AVR points to concerns about hot-spots in the reactor core and the physical integrity and quality of the fuel pebbles as constructed and handled at that time. Defects in Triso fuel construction might lead to the possibility of the release of radioactive fission products and minor actinides in the event of a serious accident. Defects would also relate to the cleanliness of the internal reactor components. The structural integrity of pebble bed Triso fuel is of particular importance, given the high-temperatures anticipated in several accident scenarios.

- Nuclear waste management has been one of the most intractable and thorny issues in the history of nuclear power. It is noteworthy, therefore, that the pebble bed concept would generate significantly greater volumes of radioactive waste than its light water reactor competitor technologies. Lyman goes so far as to calculate that the difference in waste volume is 13 times greater for a PBMR than for a PWR producing equivalent amounts of electricity [57]. PBMR proponents point out that the wastes from PBMR systems are well characterized and very easily handled as the fuel elements retain and encapsulate well the waste materials. They also add that because both the uranium density and the heat density are lower in spent pebble bed fuel than in spent light water reactor fuel, it may be possible to store and dispose of spent pebble bed fuel in closer packed arrangements than is possible with spent LWR fuel and, therefore, the difference in repository volumes required might not be as large as the critics warn. While the pebbles may go through the reactor several times (i.e. are not strictly 'once-through') they are never reprocessed. Once they leave the site the nuclear materials do not return. The fuel cycle is therefore simple and straightfor-ward.

II.6.3   Prismatic high-temperature gas-cooled reactors

The pebble bed is not the only fuel-handling concept appropriate to high-temperature gas-cooled reactors. The dominant alternative to the pebble bed approach is known as the 'prismatic' design. The Dragon Reactor Experiment (DRE) at Winfrith in the UK has been extensively reviewed by R A Simon and P S Capp [58]. The DRE was an early implementation of the prismatic approach to HTGR design. As Simon and Capp point out, the reactor had a core of overall diameter 1.08 m consisting of 37 fuel elements placed in a hexagonal array. The hexagonal array was surrounded by 30 prismatic graphite blocks, which formed the reactor's primary or 'inner live' reflector. The graphite blocks were machined on their inner faces to match the shape of the adjacent fuel element while the outer face of the graphite blocks were machined so that taken together they formed a circle of diameter 1.5 m. The reactor control rods were inserted into the prismatic reflector and not between the fuel elements. The reactor was helium-cooled with an inlet temperature of 350 °C and an outlet temperature of 750 °C. The helium flow (by mass) was 9.62 kg/s at a pressure of 20 atm and the reactor had a maximum thermal power of 21.5 MW [58].

The prismatic reflector design gives its name to an entire class of HTGRs—the prismatics. Most modern prismatic designs employ Triso fuel kernels of a type similar to, or identical to, those found in pebble bed systems. Rather than packing the Triso fuel into spherical fuel pebbles, the prismatic approach is to construct precisely manufactured fuel elements. In the DRE a standard fuel element consisted of a large graphite block with a lifting head. Machined into the graphite block were channels for six metal-clad fuel rods and a central channel for experimental studies [58]. The fuel assembly of a modern prismatic reactor will be discussed later in the context of the Japanese High-Temperature Engineering Test Reactor.

One of the most important findings from the DRE was the lack of fission product or other radioactive contamination of the helium gas. The reactor also showed a remarkable absence of corrosion [58].

The Fort St Vrain reactor, discussed earlier, was also prismatic in design, but had a completely different level of reliability and success when compared with the DRE. It was postulated earlier that this was because, while the nuclear engineering involved was robust, the mechanical and materials issues were beyond the capabilities of the time. The much more positive DRE experience in England shows, however, that prismatic HTGRs are not inevitably unreliable or unsound. One simply must be careful that the designed outlet helium temperature is compatible with the engineering adopted and that the fuel/reflector system is well constructed.

Despite what was undoubtedly a painful experience for General Atomics with the Fort St Vrain reactor, the company is still interested and keen to develop modern HTGR systems.

II.6.3.1 Japan's High-Temperature Engineering Test Reactor

The High-Temperature Engineering Test Reactor (HTTR) is the only high-temperature gas-cooled reactor in Japan. It represents the culmination of more than 20 years' research and development by the Japan Atomic Energy Research Institute (JAERI). The HTTR test reactor (figure II.6.9), constructed at Oarai Town in Ibaraki prefecture, achieved criticality on 10 November, 1998. The 30 MWth HTTR is unusual in that its motivations extend beyond improvements to electricity generation and the nuclear fuel

Figure II.6.9. HTTR building, Oarai Town, Japan (source: JAERI).

cycle. The primary motivation is a desire to obtain a flow of very high-temperature helium gas for materials and chemical research [59]. In particular the nuclear heat is being studied as a mechanism for the production of hydrogen from natural gas and water (see section I.3.4). Hydrogen is an important 'energy vector' by which nuclear power may be mobilized towards transport needs. Hydrogen is being actively researched worldwide as a vehicle fuel.

The HTTR is a high-temperature gas-cooled reactor (operating at approximately 950 °C) of the prismatic type [59]. The active core is 2.9 m tall and 2.3 m in diameter. The core consists of 30 fuel columns. Reactor control is provided by 16 control rod columns, seven of which are in the active core and nine of which are in the reflector which is also partly prismatic (12 replaceable columns).

As with most other modern HTGR systems the HTTR employs Triso fuel particles. In the HTTR case these have a UO_2 kernel with approximately 6% enrichment [59]. As is typical in modern prismatic HTGR designs the Triso fuel is sintered into fuel compacts. For the HTTR, the fuel compacts are hollow cylinders 39 mm tall and 26 mm in outer diameter. The fuel compacts are stacked in a graphite sleeve to form a fuel rod of diameter 34 mm and length 577 mm. The prismatic fuel elements are graphite blocks of width 360 mm across the flats and 580 mm in height. The prismatic blocks have vertical channels to accept the fuel rods and test samples. The configuration of an HTTR fuel element is shown in figure II.6.10.

Control of the reactor core in normal high-temperature operation relies upon the control rods. For shutdown the reflector control rods are inserted before the insertion of the control rods in the active core [59]. In an

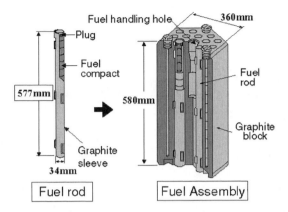

Figure II.6.10. Configuration of a prismatic fuel element of the Japanese HTTR high-temperature gas-cooled reactor (source: JAERI).

emergency shutdown (a 'scram') the gear teeth engaging the control rods are withdrawn and the control rods fall into place under gravity. A backup shutdown capability exists by inserting boron carbide/graphite pellets into holes in the control rod guide columns.

The HTTR has three cooling systems—a main cooling system (MCS), an auxiliary cooling system (ACS), and the vessel cooling system (VCS) [60]. In normal operation the entire cooling of the 30 MWth generated in the core is provided by the main cooling system. The MCS in turn has three components—a primary cooling system (PCS), a secondary helium cooling system (SHCS), and a secondary water cooling system (SWCS). The complexity and multiple redundancy of the HTTR reflects the fact that the reactor is designed to operate at the highest temperatures currently achievable and that its primary research focus is in the area of nuclear heat utilization and management.

The HTTR has two alternative standard cooling modes. One is a mode in which all of the reactor heat is managed by a pressurized water-cooling system. The other mode uses the same water-cooling system but only to extract two-thirds of the reactor heat. The other 10 MWth are handled by a helium-to-helium heat exchanger known as the intermediate heat exchanger. This heat exchanger is the highest temperature heat exchanger in operation worldwide.

HTGR systems continue to push the boundaries of engineering, and the Japanese HTTR (figure II.6.11) has not been without its technical difficulties. For instance, the HTTR was forced to shut down with problems in its gas circulator system in October 1999. The problems were fixed, however, and power-up tests were restarted a few months later in April 2000. Things progressed smoothly and on 31 March 2004 JAERI started high-temperature test operations of the HTTR. On 19 April 2004, the HTTR reached the reactor outlet coolant temperature of 950 °C and achieved its rated thermal

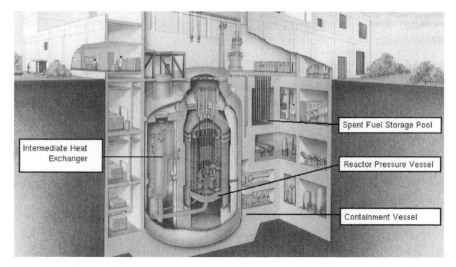

Intermediate Heat Exchanger

Spent Fuel Storage Pool

Reactor Pressure Vessel

Containment Vessel

Figure II.6.11. Layout of the JAERI high-temperature engineering test reactor (source: JAERI).

power of 30 MW [60]. It is planned to connect the hydrogen production system in 2008 [60].

One of the difficulties of HTGR technologies arises from the fact that the greatest thermodynamic efficiencies arise at the highest operating temperatures. In direct cycle systems (in which electricity generating turbines are driven by the helium passing through the reactor core) the difficulties to be overcome centre on the reliability of the turbine system in the very high-temperature gas flow. In the case of designs in which the reactor coolant circuit is separated from the secondary circuits (which may be used to drive electricity turbines) the engineering challenge is in developing heat exchangers able to operate and survive in very high-temperature conditions. It is this latter path that the HTTR addresses. As we shall see in the following section the General Atomics GT-MHR system faces the challenges involved in a direct cycle approach.

II.6.3.2 Gas Turbine–Modular Helium Reactor

The Gas Turbine–Modular Helium Reactor (GT-MHR) is a joint project of General Atomics, Minatom (and other Russian companies), Framatome ANP and Fuji Electric.

The motivations for the project are two-fold:

1. **Disposition of surplus Russian weapons plutonium.** The GT-MHR project is one strand of Russian efforts in partnership with western countries, most notably the United States, to reduce the risks of nuclear

proliferation arising from the collapse of the Soviet Union and the decline of the Russian military industrial complex. Russia chose the GT-MHR approach because: the design has attractive safety attributes, it has very high (48%) electricity production efficiencies, it achieves a higher level of plutonium burn-up than all competitors, and finally because the GT-MHR fuel cycle offers a high degree of proliferation resistance. The Russian and the American contributions of effort to the project have been substantial. The Russian contribution coordinated by Minatom includes involvement from several separate institutes including the Kurchatov Institute, while the US involvement is headed by General Atomics.

2. **The production of competitively priced electricity.** Key to the attractive economics of the GT-MHR design is the small size and compactness of the system. The GT-MHR design is striking because, essentially, the entire plant is contained within two interconnected pressure vessels, all enclosed in a concrete structure and built below ground. The design has a very small footprint and is exceptionally resilient to attack or accidental external damage. The schematic basics of the GT-MHR design are shown in figure II.6.12.

One vessel contains the prismatic reactor core cooled by helium. The helium gas leaves the reactor at an extremely high temperature (approximately $1560\,°C$). The higher the operating temperature, the higher the thermal efficiency, but also the greater the engineering challenges. The GT-MHR has the highest operating design temperature of any gas-cooled reactor (and quite possibly any reactor) currently proposed. The key engineering difficulty is the bearings for the direct (Brayton cycle) power conversion system driven by the hot helium gas stream. At the very high temperatures proposed any kind of lubricated bearing would soon prove unreliable. For this reason the GT-MHR design employs magnetic bearings. The whole power conversion system, consisting of a generator, a turbine and two compressors, is all mounted on a single shaft all contained in the second pressure vessel of the interconnected double pressure vessel design. The power conversion system is shown schematically in figure II.6.12 and in more precise detail in figure II.6.13.

Within the power conversion pressure vessel are three compact heat exchangers. One of these is a 95% effective recuperator, which recovers turbine exhaust heat and raises plant efficiency from 34 to 48% [61].

The economic advantages of the GT-MHR are further enhanced by the very small number of components needed for the power plant's construction. Most of the items involved are suitable for modular construction in factory environments.

The design team also cite other advantages, such as an emphasis on passive safety and long lead times for operator intervention in all circumstances. General Atomics expects that these safety attributes can

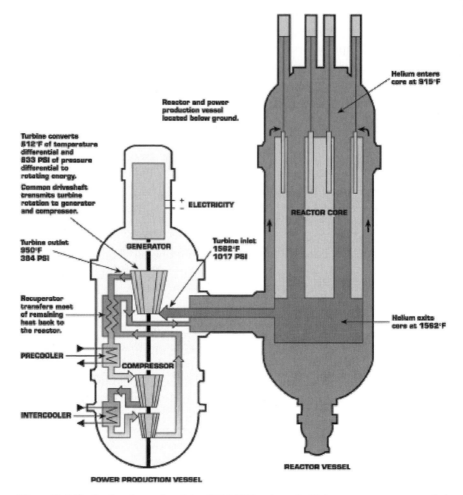

Figure II.6.12. Basic schematic of the GT-MHR prismatic high-temperature gas-cooled reactor (source: General Atomics).

lead to smoother regulatory approval and smaller (and cheaper) levels of operations and maintenance staff.

The GT-MHR reveals its Russian roots through the design consideration that the waste heat from the GT-MHR design is at an ideal temperature for use in district heating. Russia has made significant use of district heating from nuclear power ever since the beginnings of the Soviet Union civil nuclear power programme.

The GT-MHR programme is currently planning for a 600 MWth/ 293 MWe plant to be constructed at Seversk in Russia (a city known in the Soviet era as Tomsk 7). Mabrouk Methnani of the IAEA has reported

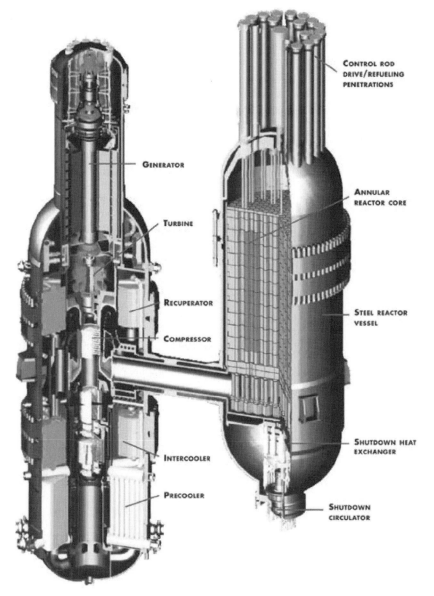

Figure II.6.13. Illustration of the interconnected double pressure vessel design of the GT-MHR.

that the GT-MHR project team is aiming to finalize the design by 2005 and to construct the first plant by 2009 [48].

On looking at the GT-MHR proposals one cannot fail to be impressed by the vision and the engineering hubris embodied in such a bold and ambitious design. One cannot fail to ask oneself, however, whether General Atomics is going to repeat the mistakes of Fort St Vrain. It is widely perceived that at Fort St Vrain General Atomics became seduced by the engineering challenges and was simply too ambitious. The GT-MHR proposal with its radical and compact design and its extremely high-temperatures could be too cutting edge for good sense. My view, however, is that things are different this time. The Russian contribution is both real and valuable. General Atomics really cannot have failed to learn the lessons from Fort St Vrain. The ambition is high but with some of the best nuclear engineers from both sides of the old Cold War divide involved, this commentator at least believes that they may just pull it off. Whatever happens, reliability will be the key.

Chapter 7

Nuclear waste burners

Nuclear power plants are not the only technological element of the nuclear fuel cycle. Several countries including France, the UK and Russia have developed nuclear fuel reprocessing technologies. Conventional PUREX reprocessing produces reusable uranium (following re-enrichment), plutonium, intermediate level wastes and high level wastes. As discussed in chapter 4, the orthodox and mainstream approach to nuclear waste management has been to propose packaging of the materials and then the disposal of them deep underground in a geological repository far from the biosphere. The encapsulation, the sealing of the repository and the various geological strata will each limit the diffusion of harmful radionuclides until such time as they have undergone radioactive decay to safe levels. Chapter 4 reviewed prospects for nuclear waste policy and addressed the role of this most troubling of nuclear power policy issues in the context of any nuclear renaissance.

In this chapter we shall consider technologies that would allow radioactive waste policy-makers to look beyond the orthodox approach of direct geological disposal. This chapter considers a possible technological modification to the back end of the nuclear fuel cycle known as partitioning and transmutation. Such an approach to the fuel cycle would build upon prior work in both reprocessing and in deep geological disposal. It is an approach designed to reduce the volume, hazard and longevity of wastes requiring deep geological disposal. It would a very expensive technological challenge, but it is one that is receiving increasing attention worldwide.

In this chapter we shall use the term transmutation to include not only the conversion of radioactive materials to stable nuclides by neutron capture, but also the fission of actinide materials, a process sometimes termed 'incineration'.

Prior to transmutation, however, the components of the wastes must be sorted and separated to ensure true benefits from the process of transmutation. This preparatory process of separating the various waste stream components is known as partitioning. The preparation of the separated components ready for later processing is known as conditioning. We shall consider some of the essential issues of partitioning in the section that follows.

II.7.1   Partitioning

In chapter 4 we considered issues in the classification of radioactive wastes. Rather than simply the classification of wastes and their conditioning as ILW and HLW, partitioning it is a more active and detailed process of chemical separation of radioactive waste materials according to their suitability for particular types of management. Some partitioned waste forms would consist of chemically pure wastes well suited to transmutation while other components would be optimized for geological disposal via the removal of particularly hazardous isotopes. It is important to acknowledge that if applied to existing high level wastes, or spent fuel, partitioning would likely be both difficult and expensive. The process of partitioning spent fuel is akin to reprocessing, although the desired output streams are somewhat different. Rather than the current reprocessing outputs discussed in chapter 4, a partitioning approach might attempt to yield the following products: uranium for re-enrichment and new fuel manufacture, a long-lived fission product waste stream (or more likely streams) amenable to transmutation, such as chemically separated technetium, iodine etc., an actinide waste stream suitable for incineration (perhaps containing both plutonium and higher actinides in a mixed state), a high level waste stream unsuitable for transmutation (containing, for instance, caesium isotopes) and the stripped metal claddings from the fuel assemblies (ILW).

As some of the partitioned waste streams would require relatively conventional management (such as that part of the high-level waste stream unsuitable for transmutation and the intermediate level waste stream), it is clear that a deep underground waste repository would be required even within a partition and transmutation (P&T)-based waste management strategy. The benefits of P&T, however, are that the radiotoxic inventory in the repository would be greatly reduced, the average half-lives of the wastes could be much shorter and the size of the repository needed could be significantly reduced. If nuclear power production is indeed to undergo a long-lasting renaissance, then this last point could be of particular importance.

Various chemical processes are proposed for the partitioning of spent fuel. The most readily available process would be to build upon the 60 year old aqueous process known as PUREX. This process has formed the basis of commercial scale nuclear fuel reprocessing over several decades. PUREX is dedicated to the separation of chemically pure plutonium and uranium from spent fuel. The process relies upon aqueous chemistry, and following the stripping of the fuel cladding consists of dissolving the spent fuel pellets in concentrated nitric acid. At the end of the PUREX reprocessing process considerable quantities of highly radioactive material remain dissolved in the acid. It is this high-level waste liquid that in the UK and France is vitrified with boron, in stainless steel containers welded shut for cooled storage and eventually for geological disposal. One minor actinide isotope in particular, americium-241, contributes significantly to the activity of the vitrified high-level waste. This isotope in particular is well suited to fast neutron based transmutation. It is poorly suited to management in a thermal neutron based process (such as would occur if it were allowed to form part of a MOX nuclear fuel) as in such circumstances some of the americium would be converted to curium-242 and 243 by neutron capture [62]. As we shall discuss later, curium has particularly troublesome isotopes, the production of which are best avoided.

As the processes of partitioning would probably be intimately connected to the science and technology of reprocessing, it is worthwhile reviewing the processes involved in a reprocessing-based fuel cycle. Figure II.7.1 illustrates schematically the closed reprocessing-based fuel cycle developed by BNFL in the UK for Magnox, AGR and LWRs. It is relevant to note that the only British commercial LWR, Sizewell B in Suffolk, does not yet have its spent fuel reprocessed, and the only LWR reprocessing work undertaken by BNFL is done under international contracts.

A more detailed insight into the various process of conventional nuclear fuel reprocessing is given in figure II.7.2.

As noted earlier, however, not all countries with advanced nuclear power programmes have adopted a reprocessing-based fuel cycle. For instance, the United States unilaterally abandoned commercial nuclear fuel reprocessing in the mid 1970s. The key basis of this decision was a concern for the weapons proliferation risk that might arise from a civil 'plutonium economy'. To this day the US retains an aversion to commercial nuclear fuel reprocessing, and most particularly to the separation of plutonium.

As the long-standing practice of reprocessing for uranium recovery and the possible future technique of spent fuel partitioning for transmutation are such inter-related activities, any US enthusiasm for P&T risks challenging that country's long established policy aversion to civil reprocessing. By way of a compromise between its concern for non-proliferation and its interest in research into new waste management techniques, the US emphasizes aspects of partitioning research that minimizes proliferation risks. Two particular consequences arise: a preference for possible UREX aqueous approaches (in which plutonium is not separated from the minor actinides such as americium) and also for non-aqueous 'pyrochemical' techniques. Pyrochemical approaches are regarded as fundamentally more proliferation-resistant than more conventional aqueous approaches with their heritage of plutonium separation [64].

Advocates of partitioning and transmutation often stress the research challenges involved. Perhaps this is because transmutation combines the

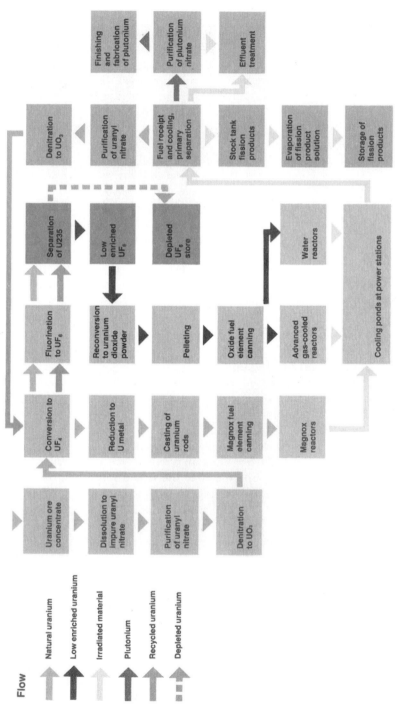

Figure II.7.1. A 1980s diagram of the UK reprocessing-based fuel cycle (source: BNFL [63]).

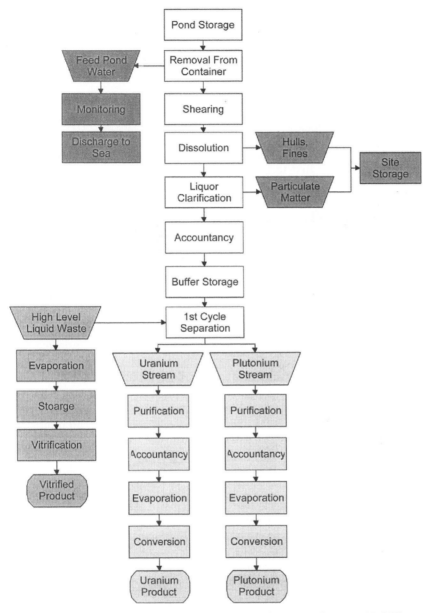

Figure II.7.2. Thermal oxide reprocessing (based on information from BNFL [63]).

ancient fascination of alchemy with the technological challenges of 'big science'. The enticing scientific challenges are not limited to the transmutation side of P&T. Many of the most interesting and challenging aspects of the problem relate to partitioning. Some of the innovation in the area of partitioning might involve approaches even more radical than the pyro-chemical 'dry' chemistry methods briefly mentioned above. Examples include electrochemical separation and magnetic separation. A wide range of waste separation techniques are discussed in a comprehensive book edited by Carleson *et al* [65]. It is perhaps useful if we review each of the candidate methods in turn.

II.7.1.1   Aqueous separation techniques

Two types of approach based upon aqueous chemistry have been investi-gated in depth. The first technique, based upon ion exchange, suffers from the need for very large plant and significant engineering difficulties, especially with highly active wastes. The second approach, based upon solvent extrac-tion, is the one that has come to lie at the heart of commercial nuclear fuel reprocessing, one of its key benefits being that at every stage in the process the nuclear materials are in liquid form [62].

The first solvent extraction process was developed at Hanford in the north-western United States in the early 1950s. The process known as 'redox' dissolved the outer fuel cladding, increasing the volume of liquid wastes involved. Furthermore, dissolving the nuclear fuel in the solvent hexone required the addition of aluminium nitrate salt in order to be able to extract the plutonium and uranium. This salting process prevented the later condensing of the active waste liquor by evaporation. The consequence of the Hanford approach was a problematic legacy of large volumes of highly active liquid waste and much physical plant [62].

In the UK efforts had concentrated on an alternative solvent known as 'butex' (di-butoxy-diethyl-ether). This allowed for the extraction of uranium and plutonium into the solvent with the addition of nitric acid only. Salting was not required and the liquor could be straightforwardly condensed via evaporation. In the UK, the fuel claddings were stripped mechanically using a die, further simplifying the processes involved in handling highly radioactive acids.

In the mid 1950s the redox and butex processes were superceded by the PUREX aqueous solvent extraction technique based around the solvent TBP (tri-n-butyl phosphate). This solvent avoided the need for salting, was chemically efficient, stable and extremely cost-effective. Since its first large scale operation at the Savannah River facility in the US in 1954 it has come to dominate large scale reprocessing. In modern reproces-sing operations the TBP solvent is diluted using odourless kerosene or 'OK'.

The US accelerator transmutation of waste (ATW) proposal from Los Alamos National Laboratory envisages a partitioning and fuel treatment based upon a UREX (or pyrochemical) approach [64]. The UREX proposal is for aqueous solvent extraction, but without the separation of plutonium from the minor actinides. The UREX process would produce purified uranium for disposal or re-use, together with separate technetium and iodine waste streams (two particularly troublesome radioactive waste materials well suited to transmutation) and a mixed actinide and fission product oxide waste stream. This final stream might be conditioned for geological disposal, or possibly even be further separated by electrometallurgical processes (see later discussion of electrochemical techniques). Such an extra step would separate the actinide materials from the fission products. The resulting actinide or transuranic (TRU) waste stream might then be fabricated into metallic forms suitable for incineration in an accelerator-driven transmutation facility. Separation of the various transuranic materials element by element is possible using what is known as a TRUEX process [66]. However, such separations are not likely to be necessary for the successful operation of a transmutation system if a fast neutron flux is used.

Despite its likely lack of direct importance to P&T the TRUEX process is noteworthy in the following respects.

The canonical form of the process, as developed at Argonne National Laboratory in the US, is a step subsequent to the conventional PUREX process. A second solvent stage is used based around the CMPO (octyl(phenyl)-*N*,*N*-dibutylcarbamoylmethylphosphine oxide). As well as separating the various minor actinides, the CMPO-based TRUEX process also greatly reduces the amount of residual uranium and plutonium in the final waste stream. Rao and Choppin reported some difficulties remaining with the CMPO-based process that required further research when they wrote in 1996 [66]. Primarily these questions related to the final separation of the radioactive materials from the solvent, the need to distinguish trivalent lanthanide fission products from trivalent transuranics, and the need to provide a high-level waste stream chemically compatible with conditioning via waste vitrification.

The need to remove lanthanide fission products is especially important as several of them (e.g. gadolinium isotopes) are strong neutron absorbers. Gruppelaar *et al* report that recently several groups worldwide have reported a set of breakthroughs in their attempts to separate trivalent americium from trivalent lanthanides [62]. At present none of the methods reported seems to be free of problematic side effects, but it seems that the difficult problem of higher actinide separation is on the verge of solution. This capability would open up a wider range of transmutation options.

The Los Alamos team are keen to explore the use of fuel treatment approaches with the highest degree of inherent proliferation resistance. They are seeking to avoid the separation of weapons-usable fissile material at all stages in their process. While the UREX proposal is consistent with such aims, it seems that pyrochemical approaches are even better suited to proliferation risk minimization.

II.7.1.2 Pyrochemical separation

Two types of pyrochemical separation have been investigated in depth: melt slagging and carbonyl refining [67].

Melt slagging involves the melting of radioactive metals and, as such, is similar in operation to the processing of liquid metals in industry. Melt slagging relies on the fact that the different components of the melt have differing tendencies to oxidize or to form silicates. As each element in turn forms an oxide or silicate, it will precipitate out as a solid at the bottom, or more usually on the surface, of the liquid metallic melt. These solid slags can then be removed relatively easily as highly pure separated wastes.

Carbonyl refining relies on the formation of gaseous metallic carbonyls. Nickel, iron and cobalt readily form carbonyls (e.g. $Fe(CO)_5$) and can be encouraged to emerge as vapours from a melt of molten liquid metallic waste. Nickel is the most susceptible to this process and Bronson and McPheeters report that nickel has been purified commercially using carbonyl technology since 1902 [67]. In some cases the necessary reactions may be problematically slow (e.g. $Fe(CO)_5$) or the resulting vapour may suffer from low volatility (e.g. $Co_2(CO)_8$ and $Co_4(CO)_{12}$).

It seems likely that the role of carbonyl refining is somewhat limited and that it is most likely to find a role as part of a broader range of pyrochemical activities.

Pyrochemical approaches need much more research work if they are to contribute to P&T on an industrial scale. They benefit, however, from the fact that it is harder to extend the methodology to include the separation of plutonium than is the case with aqueous solvent-based reprocessing, even if one aims to restrict one's concerns to UREX alone.

Looking beyond conventional aqueous techniques and pyrochemical approaches some even more radical methods have been proposed, including electrochemical approaches.

II.7.1.3 Electrochemical separation

Electrochemical separation is the use of an applied electric field to a liquid containing dissolved radioactive waste. The electric current arising from the applied field consists of a flow of electrons to one electrode and of positive ions to the other. The mobility of these ions depends upon their chemical and physical properties, and hence differs from one radioactive element to another. Differentiation between different radioactive elements is further enhanced if a semi-permeable membrane is placed between the two electrodes to inhibit the flow of ions. A key benefit of electrochemical methods is the relatively small amounts of secondary waste generated compared with conventional aqueous techniques [68].

Progress on electrochemical separation has been held back for many years by an absence of suitable electrodes and ion exchange membranes.

These materials gaps are now filled and the possibility exists for a separation technique with remarkably little secondary waste generation. Electrochemical separation typically employs aqueous electrolytes and as such is well suited to an evolutionary advancement of standard aqueous-based approaches. Electrochemical separation has been demonstrated at laboratory levels with either low concentrations or small waste volumes. Further work is required for industrial scale development of the technology [69].

II.7.1.4 Magnetic separation

Magnetic separation is a physical rather than a chemical separation technique. As such it one of the most promising of a family of physical techniques that also includes electrostatic separation. These techniques are well suited to granular radioactive material in which different grains have different radiological properties. The radioactive materials of interest are not combined together within single chemical compounds. One analogy would that physical separation would allow for a separation of salt from ground pepper if the two became mixed, but would not allow a separation of the sodium from the chlorine in the salt itself. It turns out that many of the problems associated with partitioning are amenable to physical techniques although it seems clear that more than physical techniques would be required in a P&T regime.

It is noteworthy that many of the most challenging isotopes in nuclear waste management have easily generated paramagnetic compounds. Such compounds, if physically mixed, would be well suited to separation using magnetic techniques. Table II.7.1 illustrates the magnetic susceptibility of a wide range of spent nuclear fuel-related compounds [68]. These data are presumably room temperature values. The entries with negative values indicate a 'diamagnetic' rather than a conventional paramagnetic response;

Table II.7.1. Volume magnetic susceptibility of selected compounds and elements [68].

Compound/ element	Susceptibility ($\times 10^6$) (SI units)	Compound/ element	Susceptibility ($\times 10^6$) (SI units)
FeO	7178.0	UO_3	41.0
Fe_2O_3	1479.0	CaO	−1.0
UO_2	1204.0	ZrO_2	−7.8
Cr_2O_3	844.0	MgO	−11.0
NiO	740.0	$CaCl_2$	−13.0
Am	707.0	NaCl	−14.0
Pu	636.0	CaF_2	−14.0
U	411.0	SiO_2	−14.0
PuO_2	384.0	MgF_2	−14.0
CuO	242.0	Graphite (C)	−14.0
RuO_2	107.0	Al_2O_3	−18.0

that is, the forces given to the waste components would be in an opposite direction in those cases. Rather than being a difficulty this aspect makes magnetic separation even more effective and practical.

The actinides (U, Pu, Am etc.) and their compounds have very rich magnetic properties that vary greatly with temperature. There are many exotic magnetic phases and extreme changes in magnetic response if these materials are cooled [70]. This too may be a beneficial area of research in the area of magnetic separation.

Whatever the temperature, the key to magnetic separation is that the higher the susceptibility, the easier it is to use magnetic techniques. Two particular variants of the technique are discussed in Laura Worl's work [68]. One is the separation of dry powders in a field gradient from a 1 T superconducting magnet. The other, requiring higher magnetic fields, is able to separate materials in solid, liquid or gaseous form and is based on technologies originally developed for working with clays for the paper and rubber industries [68].

II.7.2 Nuclear waste transmutation

As noted earlier, the term transmutation is used to describe two related processes. These are the conversion of radioactive materials to stable nuclides by neutron capture, and the fission of actinide materials, a process sometimes termed 'incineration'. An example of the first type of nuclear reaction is the transmutation of technetium-99. The following is an example of transmutation by neutron capture and beta decay to higher mass stable nuclides [71]:

$$^{99}\text{Tc} + \text{n} \longrightarrow \, ^{100}\text{Tc} \xrightarrow{\beta\text{-decay}} \, ^{100}\text{Ru} + \text{n} \longrightarrow \, ^{101}\text{Ru} + \text{n} \longrightarrow \, ^{102}\text{Ru (stable)}.$$

Technetium-99 is especially problematic in radioactive waste management as it is one of the most long-lived (half-life: 211 thousand years) isotopes found in radioactive wastes. Furthermore it presents a significant radiological hazard with specific activity of 0.17 curies per gram [72].

Figure II.7.3 illustrates schematically the basics of a transmutation process. In this case the core technology is an accelerator-driven system (ADS). This is consistent with the majority of transmutation research currently under way around the world, although significant interest in the use of more conventional critical reactor technologies persists. The lower left part of figure II.7.3 illustrates a conventional approach for the disposal of radioactive waste. The material is sealed in drums and placed deep underground in a large underground repository where it must remain immobilized for periods of more than 10 000 years.

Figure II.7.3, however, also shows an alternative path for that same waste. A box illustrates the atomic level process by which the individual

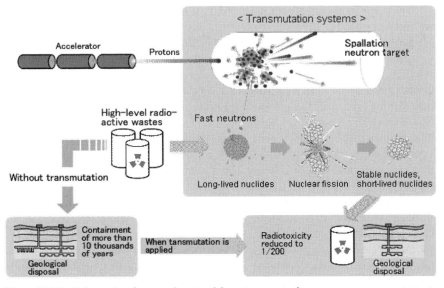

Figure II.7.3. Schematic of an accelerator-driven transmutation waste management strategy and a conventional disposal alternative without transmutation (source: J-PARC, Japan).

radioactive atoms of the waste are bombarded by fast neutrons produced when protons from a high-energy particle accelerator are fired at a heavy metal target. The process by which the fast neutron is chipped out of the heavy metal target atom by the high-energy proton is known as 'spallation', based on a technical term borrowed from geology for flaking shards from rocks such as flint. As shown in the figure, once the fast neutron strikes the radioactive waste atom it is absorbed, putting the atom into an excited state. In this case the atom is an actinide and the fast neutron causes nuclear fission. The two new nuclides produced decay rapidly to form stable non-radioactive atoms. The outputs of the process are further partitioned and the resulting radioactive waste is significantly reduced in terms of both volume and half-life. This means that, while a deep geological waste repository is inevitably required, it can be smaller and it can be optimized for a shorter design lifetime than the more classical alternative shown to the left of the figure. The shorter design lifetime is a key benefit as much of the technical uncertainty concerning long-term geological disposal of radioactive waste relates to the geological and hydrological processes at very long timescales, such as 100 000 years or more. By moving to a P&T management approach with shorter repository design lives, the emphasis of concern can shift away from geological isolation towards the materials and metallurgical properties of the immediate waste containers. This is an area of scientific research with fewer uncertainties and lower complexity.

Any move to industrial scale P&T devoted to spent fuel management will rely upon extremely efficient partitioning processes. Even with the highest probable efficiencies a real P&T facility would typically require multiple iterations of the process in order for significant improvements to be achieved. As such, the technological challenges involved in partitioning are at least as great as those in transmutation. That said, the technological challenges of transmutation are by no means negligible.

Figure II.7.3 illustrates a transmutation scheme based upon fast neutrons. This has become the dominant approach worldwide. Concerns have been expressed that transmutation could be an unacceptably slow process at fast neutron energies. In considering the transmutation rate for actinides two factors are particularly important. The first is the fission cross section which for fissile materials is indeed significantly lower at fast neutron energies. The second, however, is the neutron flux available which would typically be far higher in a fast neutron system. Fast neutrons have the added benefit over thermal neutrons that all actinide isotopes can be transmuted using fast neutrons. Koch and co-workers stress a key merit of fast neutron actinide incineration over the use of thermal neutrons when they say [73]:

> *Generally fast neutrons are more efficient than thermal neutrons, because the latter favour neutron capture processes which leads to the formation of higher actinides. For example, if the Pu is exclusively recycled in LWR, one third would not fission and hence be transmuted into minor actinides.*

For the transmutation of long-lived fission products the use of a fast neutron system is favoured because of the enormously higher neutron fluxes in such systems.

The earlier discussion of partitioning showed how, essentially, all of the range of chemical separations that may be useful for transmutation have now been demonstrated in the laboratory. Whether such partitioning methods could be implemented on an industrial scale, and with real waste streams, remains to be demonstrated in most cases. Clearly much more research is needed, but it seems evident that a full range of approaches to transmutation is open to consideration. Worldwide there are many separate investigations under way into coupled partitioning and transmutation schema. Most notably these include major projects in Japan, Europe and the United States. Each project is considering a range of options and is required to consider differing waste legacies. As such the number of permutations of possible back end fuel cycle designs are very large and it is therefore not possible to survey them all fully here. Rather we shall consider the most common concerns and those with the widest likely relevance in any nuclear renaissance.

The components of nuclear waste that are of the greatest long-term concern because of radiological toxicity are:

- the 'minor actinides' curium-242 and -245, americium-241, neptunium-237 etc. and
- the 'long lived fission products' iodine-129, caesium-135, selenium-79, radon-226 and technetium-99.

Plutonium isotopes and uranium isotopes are also of significant radiological concern, but in the context of nuclear waste partitioning and transmutation, uranium and plutonium generally form categories of their own.

A key radiological issue for several of the long-lived fission products is their high solubility in water, particularly iodine-129 and technicium-99. These risks are affected by the geological formations selected for a disposal facility. A disposal facility constructed within a geological salt dome is likely to be dry for the foreseeable future and fission product migration is likely to be minimal in all circumstances. A granite-based geology risks significant water intrusion and careful planning of multiple barriers is required to ensure that soluble fission products are not transported by the local hydrology. In this context the balance of risk between natural evolutionary scenarios and human intrusion scenarios varies with repository geology. Gruppelaar and colleagues comment that, if expressed simply, long-lived mobile nuclides, such as fission products, dominate both normal evolution scenarios and human intrusion scenarios in clay and granite. In salt, the shorter-lived radiotoxic actinides dominate (via early human intrusion) as both actinides and fission products are naturally immobile in salt [74].

II.7.2.1   Plutonium and the minor actinides

Good partitioning is essential for the efficient P&T-based management of spent fuel. In chapter 4 we considered the importance of the 'what is waste?' question. It is important to note that currently in the UK spent nuclear fuel is not a waste and therefore in this author's opinion some of the important questions concerning partitioning for waste management have failed to figure appropriately in the UK radioactive waste policy debate. Ironically the other key material that has historically failed to register properly with British policy makers has been separated civil plutonium, approximately 100 tonnes of which are currently stored in dioxide powder form [75]. The irony is that while spent fuel requires much work for it to be prepared for transmutation, separated plutonium dioxide is already well placed for preparation as a transmutation (incineration) fuel.

In the following section we shall consider in detail the transmutation of problematic radioactive materials including separated plutonium. While separated plutonium from PUREX reprocessing is extremely pure and well characterized when measured against the usual waste streams considered for transmutation, there is however a problem. The isotope plutonium-241

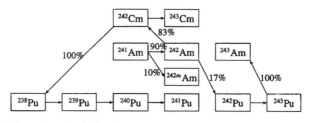

Figure II.7.4. Flow chart depicting the main transmutation response of americium-241 to a thermal neutron flux [62].

comprises 9.9% of the plutonium in UK AGR spent fuel, and 14.7% in LWR spent fuel [75]. This plutonium-241 component decays quite rapidly (half-life 14 years) to form americium-241 (half-life 433 years) [76]. As mentioned earlier, this americium isotope is particularly problematic if placed in a thermal neutron flux (e.g. in an LWR MOX fuel cycle) as curium isotopes result. For this reason americium removal would need to be a significant, expensive and troublesome part of the management of old plutonium in conventional reactor fuel cycles. This has led the UK to start considering actively the question as to what part of the national plutonium inventory might be 'surplus' and hence perhaps designated as waste, as the House of Lords has suggested [77]. If, rather than being incorporated into fuels for conventional thermal neutron reactors, the old plutonium could be incinerated in a fast neutron spectrum then it would not matter if even significant amounts of americium were present. The problem of placing americium isotopes in a thermal neutron flux is illustrated by figures II.7.4 and II.7.5 from H Gruppelaar *et al* [62]. These ideas are also discussed in a paper by R J Konings *et al* [78].

Figure II.7.4 shows that 90% of the americium-241 undergoes neutron capture to become americium-242. Of this, 83% undergoes beta decay to become curium-242. This in turn can either undergo neutron capture in the thermal neutron flux to become curium-243 or undergo alpha decay to form plutonium-238. In a thermal neutron flux plutonium-238 can undergo successive neutron captures until it becomes the relatively short-lived (half-life 14.4 years) and hence highly radioactive isotope plutonium-241. Along

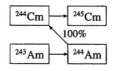

Figure II.7.5. Flow chart depicting the main transmutation response of americium-243 to a thermal neutron flux [62].

the way, however, the sequence involves the fissile isotopes plutonium-239 and plutonium-241. Much of the plutonium therefore fissions as it passes along the sequence shown in the bottom-left portion of the figure. 17% of the americium-242 undergoes beta+ decay to form plutonium-242 which in turn captures a neutron to become plutonium-243, itself undergoing beta decay to form americium-243. In summary the four consequences of placing americium-241 in a thermal neutron flux are fission and the creation of curium-243, plutonium-241 and americium-243.

Americium-243 is itself a material that one wants to keep away from thermal neutron fluxes. Figure II.7.5 shows how a succession of neutron capture, beta decay and neutron capture yields an output of curium-245.

Curium is an element that one seeks to exclude from all nuclear fuel cycles partly as a consequence of its extreme radiological hazards (which means it is expensive to handle in significant quantities [78]). Importantly some curium isotopes are spontaneous neutron emitters, a factor that can make criticality management more difficult. The significant problems of curium management prompt the conclusion that wherever possible americium isotopes should be kept away from thermal neutron fluxes.

Clearly it has been inevitable that some americium would be exposed to such fluxes in a conventional thermal reactor fuel cycle. The build up of americium in conventional spent fuel, however, is such that it and the resulting small fraction of curium produced are held together in the high-level waste stream. The curium component to the waste is a significant contributor to the minor actinide radioactivity in the high-level waste, but is regarded as no more than a part of the wider HLW problem. In seeking to partition such high-level wastes the curium problem would become more apparent. As it became apparent, however, the option of transmutation rather than geological disposal would probably become more attractive.

The arguments above, however, provide key reasons for excluding americium from the mixed-oxide fuel cycle. Americium builds up slowly in old civil plutonium as a result of radioactive decay. Disposing of old civil plutonium via a MOX fuel cycle therefore requires extra work to remove the americium prior to fuel fabrication. Americium removal is not only motivated by concerns for curium production, but americium-241 removal also makes MOX fuel manufacture more straightforward. This is because the americium isotope is a major contributor to the radioactivity of older separated civil plutonium. The more radioactive the plutonium the more burdensome the measures required for its safe handling. While the isotopic removal of highly radioactive plutonium-241 from separated civil plutonium would be prohibitively difficult, the chemical removal of americium-241 from older separated plutonium is, at last, possible. The benefits and drawbacks of americium-241 in older separated civil plutonium will be key factors if the UK is to consider formally whether some of its separated civil plutonium might indeed usefully be regarded as 'surplus'.

Gruppelaar and co-workers propose that separated curium be managed in one of two ways [79]. Either the curium should be left for approximately 100 years to allow it to decay to plutonium, or it might be mixed with americium isotopes and fabricated into an inert matrix specifically designed for fast neutron transmutation.

The problems associated with placing americium isotopes in a thermal neutron flux together with issues concerning the efficient management of long-lived fission products lead one to conclude that, for general utility, technologies based upon fast neutron fluxes are the best choice for nuclear waste transmutation. Technologies of transmutation are discussed further in the following section.

Unless significant changes are made to the UK nuclear industry (e.g. the development of AGR MOX or the construction of a new fleet of nuclear renaissance LWRs) technical, commercial and political pressures will ensure that MOX plays only a small role in addressing the UK's plutonium problem. It is also challenging to British orthodoxy that plutonium, once believed to be a valuable asset, might be recombined with the high-level waste from which it was so expensively removed in the PUREX process. Even if the UK could accept such a reverse step, it is most unlikely that there would be sufficient HLW in a form suitable for mixing with the surplus plutonium. The UK's unique plutonium legacy is a compelling motivation for that country to explore transmutation technologies for plutonium incineration. The British classification of radioactive wastes has, historically, excluded separated plutonium and this is one reason why partitioning and transmutation has been viewed with little enthusiasm by the British nuclear establishment.

It is important to note that the fission cross-sections for plutonium at fast neutron energies are very small. By itself this would cause fast neutron approaches to transmutation to be a very slow process indeed were it not for the fact that far higher neutron fluxes can be achieved in fast neutron systems. The benefit of fast neutrons is that the fission to capture cross-section ratio is higher at higher neutron energies, which means that if fast neutrons were to be used, fewer problematic isotopes of americium and curium would be produced.

II.7.2.2   Transmutation of long-lived fission products

As discussed above, long-lived fission products are a significant challenge for geological radioactive waste disposal. Only a few of the radioactive fission products produced in thermal fission reactors represent a concern for the long-term. Most have decayed within a few decades of their production. Three fission products in particular are of widespread concern because of their very long half-lives, their high radio-toxicity and their solubility in groundwater. These are technicium-99, caesium-135 and iodine-129. These particular isotopes have differing attributes when considered for partitioning and transmutation. The following discussion is informed by the material provided by Gruppelaar and co-workers in section 3.5 of their book *Advanced Technologies for the Reduction of Nuclear Waste* [62].

Technetium-99

As noted earlier, Technetium-99 can be converted to stable ruthenium isotopes by a process of single neutron capture [71]:

$$^{99}\text{Tc} + \text{n} \longrightarrow {}^{100}\text{Tc} \xrightarrow{\beta\text{-decay}} {}^{100}\text{Ru} + \text{n} \longrightarrow {}^{101}\text{Ru} + \text{n} \longrightarrow {}^{102}\text{Ru} \text{ (stable).}$$

While ruthenium-100 is normally stable, it too can undergo neutron capture in a neutron flux. The likely outcome of placing a technetium-99 target in a neutron beam would be to generate a spread of ruthenium isotopes from ruthenium-100 to ruthenium-102. If need be, these are straightforwardly separated from any residual technetium-99 by chemical means.

Technetium-99 has a high neutron capture cross section (19 barn) in the thermal energy range. However, the transmutation half-life for technetium-99 at these energies (i.e. using neutrons from an LWR with conventional UO_2 fuel) is approximately 39 years, making the process impractically slow [62].

Significantly improved transmutation rates can be obtained by using neutrons of slightly higher energy (in the 'epithermal' range) as there are resonances in the neutron absorption cross section at such energies (see figure II.7.6). These resonances provide for 50-fold increases in absorption

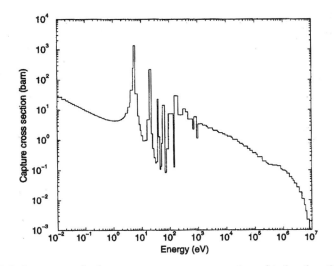

Figure II.7.6. Resonances in the neutron capture cross section of technetium-99 [62].

cross-section at particular precise energies. Epithermal neutrons suitable for resonance crossing transmutation are available from accelerator-driven systems (ADS).

Consideration of ADS approaches to the transmutation of technetium-99 has led to the proposal from Carlo Rubbia's ADS group at CERN that the technique known as 'adiabatic resonance crossing' may be used to good effect in the case of technetium-99 [62]. While the resonances may seem at first sight to be very sharp in energy, when compared with the amount of energy lost by an epithermal neutron in a single moderating collision, they are very wide. If epithermal neutrons of energy slightly higher than the resonance are allowed to thermalize within a technetium-99 target then, as they decelerate from the initial high epithermal energy to the final thermal energy, they must inevitably cross the absorption resonances. As that resonance is much wider than the energy loss per collision then each neutron will inevitably make many collisions in the resonance energy range as they slow down. There is therefore a very good chance that neutron capture will occur. Experiments at CERN have already demonstrated the possibility of the adiabatic resonance crossing approach in speeding up the transmutation of technetium-99.

Caesium isotopes

Caesium isotopes are a topic of key concern for radioactive waste management. As distinct from the case of technetium, caesium is generated in nuclear fuel use in several isotopic forms [62]. The most abundant isotope is caesium-137, which is highly radioactive with a half-life of 30 years [73]. Caesium-135 is more problematic in radioactive waste disposal owing to its longer half-life (2.3 million years). Caesium's problematic status amongst the long-lived fission products (LLFP) arises partly as a consequence of its susceptibility to biological take-up. This is particularly important given the risks associated with the fact that caesium is readily soluble in groundwater. Groundwater intrusion is a significant consideration in northern European repository locations with granite geology. Other groundwater soluble isotopes include the LLFP iodine-129 and the activation product carbon-14. The conventional European idea for the management of such water-soluble LLFPs is the 'multi-barrier concept' of deep geological disposal (see section I.4.1.1). This relies on robust waste containers and repository backfill designed to ensure that the wastes are isolated and dry until long after the problematic LLFPs have safely decayed away to negligible levels. In the UK reprocessing-based fuel cycle, most of the LLFPs form part of the high-level wastes that are vitrified with boron. The sealed stainless steel flasks containing the vitrified high-level waste are currently surface stored under active management while the radiological hazard is allowed to decay and the self-heating property of the waste diminishes.

The clear importance of the caesium isotopes to the radiological hazards of spent nuclear fuels and nuclear wastes would suggest that caesium should be addressed specially in any transmutation scheme designed to permanently eliminate radiological hazards. Unfortunately, however, caesium it is not amenable to management using techniques of partitioning and transmutation. The principal reason relates to the inevitable distribution of a range of caesium isotopes in the waste. The benefits arising from any attempt to transmute away the harmful caesium isotopes would be lost as a result of the creation of new harmful isotopes following neutron capture by the previously innocuous caesium isotopes. Partitioning is a primarily chemical process, and as such, it is not able, in any remotely cost effective way, to separate one caesium isotope from another. The continuing threat that such partitioned and conditioned caesium in the radioactive waste inventory would pose is one of the reasons that any partitioning and transmutation based waste management scheme would inevitably need a deep geological repository, albeit of smaller size and shorter design lifetime than would be required without partitioning and transmutation.

II.7.3   Epithermal and fast neutron transmutation technologies

II.7.3.1   Fast critical reactors

Nuclear reactors based upon fast neutron fission were regarded by many in the 1960s and the 1970s as the vital technological step to ensure the long-term sustainability of nuclear power. During those years energy policy-makers were troubled by, as they perceived it, the early depletion of crude oil reserves. Their forward predictions showed a continued expansion of conventional thermal nuclear power plants with uranium-based fuel cycles. This led to a second resource insecurity that, in the long term, uranium ores would be depleted as a source of uranium-235 nuclear fuel. The fast reactor concept opened up a much wider range of possibilities, as at fast neutron energies, a wider range of nuclear processes are possible, including, for instance, the transmutation of fertile uranium-238 into fissile plutonium-239. While the term 'fast' applies to the neutron energies involved, it is the aspect of the conversion of abundant uranium-238 into plutonium-239 that, in those years, led to the insertion of the word 'breeder' into the title 'fast breeder reactor'. The ability of such reactors to produce (or 'breed') their own fuel (e.g. plutonium-239) from otherwise useless materials (e.g. uranium-238) seemed to many to be the long-term solution to the anticipated depletion of the world's uranium-235.

Many countries had fast breeder reactor research programmes including the UK (at Dounreay in Scotland), France (in the Rhône valley), Japan (near Tsuruga Bay on the western coast of Honshu Island) and the then Soviet Union (at Beloyarsk in the Urals of central Russia). Despite its position as the first country in the world to successfully develop the fast reactor technology in the early 1950s, the United States has long had reservations, on grounds of nuclear proliferation prevention, about the fast breeder concept. Its aversion is not total, however, and it is currently actively considering fast reactor options within the Generation IV programme (see chapter 8).

Through the 1980s the original motivation for the fast breeder technology largely dissipated. Rather than rising inexorably, uranium prices fell in the late 1970s and remained far lower than had been predicted (see figure I.4.5). Technically fast reactors proved to be far more challenging to build and operate than had been expected. The extremely compact cores and the large amounts of heat to be removed led to designs based upon liquid metal coolants. In several countries the initial wave of research reactors was followed by the development of larger demonstration reactors. In the UK, the government of Margaret Thatcher found the arguments in favour of the fast reactor programme unconvincing and a decision was made in 1994 to wind down the programme. In France the Super-Phenix prototype reactor never lived up to its promise and is currently shut down as a result of technical and financial difficulties, while in Japan the Monju commercial scale fast reactor faces a difficult future following the embarrassing cover-up of a coolant fire in December 1995. India and China both retain fast reactor ambitions with designs at various stages of completion.

In Russia enthusiasm for fast reactor technology continued and was tempered only by the collapse of the Soviet system and the Russian economy in the early 1990s. Despite appalling economic difficulties the Russian Federation's Ministry for Atomic Energy (Minatom) kept alive an enthusiasm for the fast reactor concept. While uranium ores have remained plentiful and no threat of depletion is on the horizon a new policy driver has emerged in recent years that has encouraged Minatom to retain an enthusiasm for the fast reactor.

While fast reactors were originally sold for their ability to breed new fissile plutonium, they are now advertized as a key technology for the elimination of unwanted surplus plutonium. A technology once touted as the 'fast breeder reactor' is now sold as a 'nuclear waste burner'. Russian enthusiasm has also been kept alive by a design, never built, of a fast reactor that Minatom believes has enormous promise—the BREST concept and design. In March 2003 Minatom was reorganized as part of a shake up of government ministries in the Russian Federation. Its civil activities were passed to a new Federal Atomic Energy Agency (FAEA), while its military functions were passed to the Defence Ministry. It is not the purpose of this book to review the history of nuclear power. Rather we aim to consider technologies of greatest future promise. Of all the Russian technologies, it is the BREST fast reactor design that we shall consider in detail.

II.7.3.2 The BREST fast reactor

The BREST fast reactor concept builds upon a series of successful fast reactor projects in Russia. This experience culminated in 1980 with the completion of a third 600 MWe fast reactor unit at Beloyarsk. This BN-600 reactor represented a significant push forward in the engineering of liquid metal cooled reactors [80]. Current Russian fast reactors use liquid sodium metal as a coolant, but this is problematic owing to the high chemical reactivity of the coolant which combusts spontaneously if in contact with either air or water. The choice of sodium coolant is optimal for the breeding of plutonium in the now-outdated concept of a fast breeder reactor. Recently the larger BN-800 fast reactor project was restarted following its suspension in the 1980s following the Chernobyl disaster in 1986.

Looking beyond the BN-800, which is due to be operational by 2010, the Russians are proposing the BREST fast reactor. The Russians have adjusted their fast reactor designs to solve the problem of plutonium production in the BREST design by replacing the usual uranium dioxide fuel with mononitride (UN-PuN) fuel [81]. This increases the density of the fuel. One problem with the use of nitride fuels in fast reactors is the accumulation of biologically harmful carbon-14 in a nuclear reaction in which naturally abundant nitrogen-14 has a proton transformed into a neutron (with the emission of a fast positron) yielding carbon-14. This problem is avoided if the fast reactor mononitride fuels are manufactured with high fractions of the rarer nitrogen-15 isotope. Adamov *et al* report that isotopic enrichment of nitrogen of up to 90–99% and the later entrapment of 90–99% of the carbon in the spent fuel during reprocessing would be sufficient to solve the carbon-14 radioactive waste problem [82].

Another step towards improving the BREST design is that rather than being cooled by liquid sodium it will use a heavy metal coolant, most probably liquid lead. The Soviet Navy had much experience of heavy metal reactor coolants, having used lead–bismuth coolant in eight nuclear powered vessels. The BREST design avoids the incorporation of bismuth (a scarce and expensive material). One difficulty in moving to lead is that the melting point of pure lead is significantly higher than that of eutectic lead–bismuth mixtures.

An emphasis on proliferation resistance is important for the Russian FAEA (Minatom), if it is to be able to partner with western (and particularly American) organizations in the development of the concept. The BREST design has the possibility of plutonium breeding (breeding ratios of 1:1.1 are being considered) [81]. Despite this, it would seem that the BREST fuel cycle has the potential to be more proliferation resistant than the light water reactor fuel cycle preferred in the United States. Key proliferation-resistant steps in this fast reactor concept include the absence of any additional uranium blankets. Such blankets were ubiquitous in early fast breeder proposals as a way of generating extra plutonium by placing uranium in the path of fast neutrons emerging from the reactor active zone. Furthermore the extremely tight criticality margin of the reactor means that it would be effectively impossible to substitute a conventional fuel element with a substitute dummy fuel element designed specifically for plutonium production [81]. In addition, the BREST concept has the very powerful possibility that its entire fuel cycle can operate without uranium enrichment technologies and with no need for plutonium separation [83].

While the Russian government stresses the proliferation resistance of the BREST concept, and the role that the BREST design can play in the disposition of surplus plutonium (by operating as an actinide burner), many United States experts retain a distrust of fast reactor technology on proliferation prevention grounds. This difference of attitude is highlighted in an article by Charles Diggis [84]. This article discusses a United States Department of Energy agreement with Minatom signed at a summit between President George W Bush and Russian President Vladimir Putin. The agreement was to drive forward US–Russian cooperation on the disposal of surplus Russian weapons-usable fissile materials. Almost immediately Minatom officials were trumpeting the agreement as the basis on which to take forward the BREST concept while the US negotiators commented that they had never even heard of the BREST fast reactor proposal.

The BREST project is to start with a demonstration 300 MWe unit to be constructed at Beloyarsk by NIKIET (the Moscow Scientific Research and Construction Institute) with plans for it to be operational by 2010. The BREST 300 demonstration unit is to be followed by a commercial 1200 MWe unit by 2020. This proposed plant rating of 1200 MWe is a consequence of the maximum size of suitable turbines currently available in Russia [82]. The outline design of the BREST 1200 is shown in figure II.7.7. The outline design shows how the BREST concept is based around a coolant pool. Liquid lead coolant is poured into a shaped pit of heat-isolating concrete in which is located the core active zone, the steam generator, the pumps and other servicing systems. The concrete walls of the pool are cooled by air circulating in embedded pipes and in this way the concrete is kept below 100 °C. This means that the lead at the walls of the pool will be in the solid phase and only lead closer to the active zone is liquid and circulating [82]. The air-cooled concrete has a safety role in the event of a worst-case accident. The design is such that, if residual heat removal were required in the absence of the primary lead cooling system, then the air cooled concrete wall would not rise in temperature above 800 °C, which is well within the structural limits of this barrier. In the further unlikely event that the air-cooling channels were to become blocked then cooling in the channels can

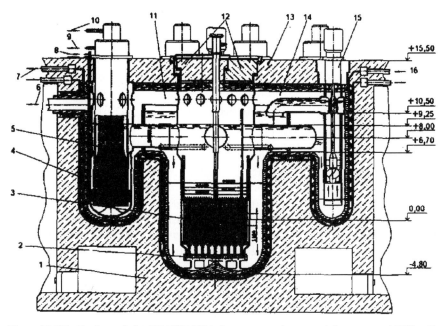

Figure II.7.7. Design of the BREST-1200 fast reactor (source: Adamov *et al* [82]). (1) Concrete vault wall, (2) support grid, (3) core, (4) thermal insulation, (5) steam generator, (6) steam discharge into pressure suppression pool, (7) air into concrete cooling system, (8) air into cooldown system, (9) steam outlet, (10) feedwater inlet, (11) pressure chamber, (12) rotating plugs, (13) upper plate, (14) gas volume, (15) circulation pump and (16) air.

be provided by the evaporation of water located in a tank at the top of the concrete pool wall. In normal operations this water tank is used to hold spent fuel removed from the reactor.

In addition to the proliferation resistance considerations discussed earlier, the Russian FAEA (Minatom) stresses the viability of the BREST concept against the key criteria of safety, economics and environmental impact.

The Russian FAEA analysis of design basis (and beyond design basis) accidents has shown that, short of complete obliteration of the reactor system by external action, the reactor system is resistant to prompt runaway, to loss of coolant, to fire, to hydrogen explosion and to fuel disintegration [81]. In all such accident scenarios radioactive release is predicted to be thousands of times lower than that which occurred in the Chernobyl accident of 1986. The Russian FAEA is confident that even in worst-case accident scenarios no evacuation of people living locally would be required.

In discussing reactor safety, the FAEA is keen that insurance assessments should consider actual reactor operation experience with risks in the range 10^{-3}–10^{-4} per year, rather than the theoretical risks of 10^{-6}–10^{-7}

Table II.7.2. Economic performance comparison of 600 MWe Russian reactor technologies [81].

| Indicator | Value, VVER | % for technologies | |
		BN	BREST
Cost of electricity	100 (benchmark)	124	93
Comprising:			
Fuel cost	23	18	17
Capital investments	54	76	54
Operations	23	30	22

that emerge from conventional nuclear industry probabilistic risk assessments [81]. The Russian FAEA believes that it is important that the nuclear industry can now point to 10^4 years of actual reactor operations experience worldwide rather than relying on demonstrably over-optimistic theoretical predictions by experts.

One aspect emerging from the Minatom safety analysis was the need to ensure site security. This is especially interesting coming as it did before the events of 11 September 2001. According to Adamov and co-workers the BREST design requires political action to prevent the possibility of a sabotage in which large positive reactivity is rapidly inserted into the active zone. While the BREST design is apparently resistant to accidental developments there are clearly concerns regarding its resilience in the face of malicious intent [81].

Traditionally economics has been one of the two show stoppers for fast reactor ambitions, the other being technical reliability. Minatom/FAEA has performed a detailed assessment of a 600 MWe BREST unit and compared it with the older Russian BN fast reactor design and with the Russian VVER pressurized water reactor. The economic analysis is presented in table II.7.2 and may be summarized by the statement that the BREST technology has similar economics to the VVER 1000 design, but is roughly 40% cheaper than the BN fast reactor design. A major source of cost savings compared with the BN design is the avoidance in the BREST design of the use of liquid sodium (with its particular hazards and challenges).

In an environmental context Minatom/FAEA has tended to the view that if a nuclear fuel cycle could be developed that yielded wastes no more hazardous than the ores mined in order to produce the nuclear fuels, then the resulting sense of a natural balance would be a key step in winning greater public acceptance of nuclear power.

The BREST reactor combined with appropriate reprocessing seems able to achieve the FAEA grand ambition of a sustainable nuclear fuel cycle with no increase in radiological burden being passed to future generations.

It is in this respect that the attributes of the BREST concept as a nuclear waste burner come to the fore. One step identified by the FAEA is the need for americium and curium recycling. The BREST fast reactor has the capacity to burn such isotopes (fast neutron fission) and to help transmute the legacy wastes of the conventional Russian nuclear power programme. Minatom/FAEA has also identified the need to co-extract long-lived radium-226 and thorium-230 isotopes for special handling. The bulk of the residual radiological burden would consist of uranium emerging from reprocessing. Minatom suggests that this uranium should all be regarded as fuel for future power generation (all isotopes being useful in a fast reactor fuel cycle).

While the BREST reactor concept is already capable of utilizing surplus military plutonium as a fuel, the more demanding step of using it as a burner for partitioned wastes from conventional nuclear power generation would require further research. One difficulty would be the need to operate in a critical system with a very tight margin for criticality. A related concern is whether the wastes should be kept separate from the reactor fuel (heterogeneous transmutation) or incorporated in the main fuel rods (homogeneous transmutation).

These concerns have been discussed further by Adamov *et al* [81]. Adamov and his colleagues from Minatom/FAEA and the Research and Development Institute of Power Engineering have very usefully addressed some of the central concerns expressed about the BREST design [81], but it is perhaps useful to mention a few of the central issues here.

First, concern has been expressed that molten lead will prove to be excessively corrosive, but Adamov *et al* report that, for instance, pump impellers can be made from temperature-resistant steels and be subject to regular maintenance and replacement. The flow rates of lead are relatively low and therefore no special stresses are expected on the pumps.

A second concern is that high temperature lead is combustible so as to generate difficulties similar to the handling of sodium, but Adamov *et al* make clear that this fear is not consistent with experimental studies of liquid lead up to 1200 °C. A third concern has been expressed that lead might freeze in the pumps and other key systems. It is reported that even in shutdown mode the decay heat on the core will be sufficient to keep the lead pool liquid in all key areas. In fact the solidified lead at the boundaries of the pool can be useful in sealing small cracks in the concrete lining of the pool.

Perhaps the most challenging issue is the very tight reactivity margin required for operation of the BREST design ($\Delta K_{\text{eff}} \approx 0.0035$) at all times in the operation of the core. This, Adamov and co-workers concede, is a stringent test and further design optimization and analysis is required to improve confidence in the routine operability of the concept. It is partly for this reason that a demonstrator scale system of rating 300 MWe is proposed.

II.7.3.3   Accelerator-driven systems

In recent years one particular technological approach has come to dominate discussion of the transmutation of nuclear wastes—the use of accelerator-driven systems (ADS). Originally proposed by Charles Bowman [85], the concept has been given new impetus on the one hand by the advocacy of Physics Nobel Laureate Carlo Rubbia of the CERN laboratory in Switzerland, and on the other hand by significant improvements in accelerator technology.

The basic foundations of the concept are as illustrated in figure II.7.8. This schematic illustrates the flow of electrical power and the flow of nuclear waste materials through an ADS system. The key to the ADS concept is the production of a high-energy, high-current proton beam. Such a beam may be produced from a linear accelerator or from a cyclotron, as shown schematically in figure II.7.8. The electrical power to drive the accelerator can be supplied from the operation of the transmutation assembly itself. The extension of the motivation of the ADS concept to include the economic generation of electricity is often associated with the design concept advocated by Carlo Rubbia and known as the 'Energy Amplifier' (see figure II.7.8).

An accelerator-driven system would be a substantial piece of physical infrastructure. Typically in such a facility hydrogen gas would be ionized and negative hydrogen ions would be accelerated into a thin metal foil that

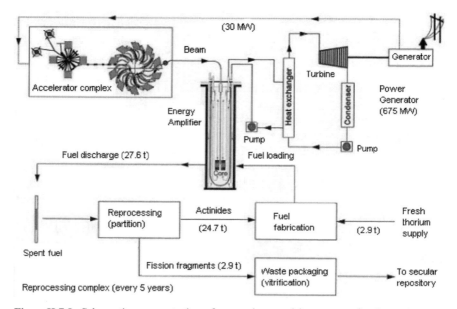

Figure II.7.8. Schematic representation of an accelerator-driven system for the transmutation of nuclear wastes. This figure illustrates the position of the ADS transmutation facility in the wider partitioning and transmutation process [86].

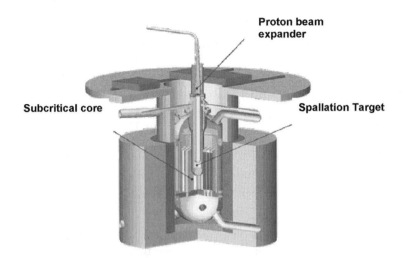

Proton beam expander

Subcritical core

Spallation Target

Figure II.7.9. Schematic representation of the core of an accelerator-driven nuclear waste transmutation facility (source: CEA, Paris [87]).

would strip the electrons from the ions leaving bare protons. These in turn would be accelerated to produce a beam with an energy of approximately 1 GeV and a current of at least 50 mA. In most ADS schema the beam is directed towards the transmutation core and turned downwards to strike the target in a vertical path. This is best illustrated in figure II.7.9.

The proton beams as produced from conventional accelerators are too narrow to be useful for bulk transmutation. There is therefore a need to broaden the proton beam magnetically so that it has a cross-sectional diameter of approximately 10 cm.

The accelerator and the beam pipes are held under vacuum to avoid degradation (by scattering from gas atoms) of the high-energy proton beam. Within the transmutation assembly the beam must pass through a thin metal window at the end of the beam pipe. The integrity of this window is fundamental to the maintenance of the high vacuum in the beam pipe and the accelerator. The beam window is one of the most critical components in any ADS for nuclear waste transmutation. While the beam currents are very high by the usual standards of accelerator beam windows, the beam size is also very large. The consequence is that the current density is unlikely to be much higher than is usual in present state-of-the-art accelerators. Nevertheless the window will need to dissipate a great deal of heat and to be able to cope with thermal shocks when the beam is turned on and off. The size of the beam window is a particular challenge as the centre of the beam will be some distance from the window cooling circuit located at the edge of the window.

Once the proton beam has passed through the beam window in most implementations it is planned that it strikes a heavy metal target where it will spall neutrons from the nuclei of the heavy metal. There is much current interest in liquid heavy-metal spallation targets. One prime candidate is the use of lead–bismuth eutectics pioneered by the Soviet Navy as reactor coolant materials during the Cold War. The term 'eutectic' means the alloy ratio of lead to bismuth is chosen such that the mixture has the lowest melting point possible. This choice allows for a liquid heavy-metal target at the lowest possible operating temperature (cf. earlier discussion of the BREST fast reactor, section II.7.3.2).

The neutrons produced by spallation techniques have a wide energy distribution in the fast and epithermal neutron energy ranges. Much of the worldwide experience in the generation of ADS spallation neutrons comes from their use in condensed matter physics. Most of these applications involve neutron diffraction and require neutrons in the thermal energy range. Such thermalization requires neutron moderation. In the case of ADS neutron spallation for transmutation, however, no such moderation is definitely required as the raw fast neutron spectrum could be well suited to the task in hand. Moderation to the epithermal range may, however, be desirable.

The importance of the beam window for system reliability was stressed earlier. In some sense one of the accidents waiting to happen in an ADS system is that the window ruptures during operations. In such an eventuality the accelerator would immediately trip out on detection of a loss of high vacuum. The lead–bismuth eutectic target material would flow through the crack in the window, at which point it would rapidly cool and solidify. The broken window and attached lead–bismuth alloy might then be removed relatively easily. One risk is that the inside of the beam-pipe might be coated with solidified droplets of lead–bismuth. The residual vapour pressure of such a material may be of sufficient concern to motivate the special cleaning of the inside of the beam-pipe if need be. In any event the failure of the beam-pipe window should not be catastrophic or prevent the rapid return to operations of the facility. More radical approaches based upon windowless designs are also being researched.

During the past ten years a substantial literature has accumulated concerning the use of ADS technologies for the fast, or epithermal, neutron transmutation of spent nuclear fuel and nuclear wastes. Three regions of the world in particular are taking these ideas forward as parallel projects. For an excellent overview of these international developments the reader is referred to Waclaw Gudowski's 1999 paper in *Nuclear Physics A* [88]. For an overview of the physics of accelerator-driven systems, the reader is referred to the recent book on the topic by Nifenecker and co-workers [89].

In Europe, activity includes a series of national projects. Prominent among these is the French experimental research programme in Grenoble.

The reactor physics group at the Laboratory for Subatomic Physics and Cosmology (LPSC) of the French IN2P3 institute has for many years undertaken experiments and computational simulations of ADS hybrid reactors and waste transmutation systems [90]. This French activity is driven by the technology challenges and the scientific uncertainties, the main policy issues in France having been settled by French national legislation in 1991. The French law requires that a public report be prepared for 2006 on three core topics: research into waste partitioning and transmutation, evaluation of options concerning retrievable and non-retrievable geological disposal, and a study of waste immobilization and surface storage [91]. This high-level statutory requirement for P&T research is relatively unusual and contrasts starkly with UK attitudes where traditionally there has been a unidirectional push towards phased geological disposal and some reluctance amongst the British nuclear industry to consider extensively other options.

While individual European national programmes have much to contribute, it is the role of international European collaboration that is particularly important in this extremely expensive and complex area. The Framework Programmes of the European Union Euratom Treaty and the independent multilateral European Centre for Nuclear Research (CERN) stand out for the scale of their contribution to the science and technology of P&T. Related European technology policy developments in the area of P&T include the April 2001 European Roadmap for Developing ADS for Nuclear Waste Incineration [92]. At CERN the key experimental step in transmutation physics has been the commissioning in 2001 of the neutron time-of-flight (nTOF) facility. This produces spallation neutrons from a lead target and allows for highly accurate measurements and experiments needed for the study of key aspects of the transmutation process (such as fast neutron absorption cross-sections). The nTOF facility is a direct precursor of Carlo Rubbia's ambitious proposal for an Energy Amplifier.

As figure II.7.3 makes clear, accelerator-driven transmutation of radioactive wastes does not eliminate the need for a geological waste repository. If P&T lives up to its potential, however, the repositories required would be smaller and easier to design and engineer. A key advantage is that the radioactive lifetime of the wastes would be significantly shorter, so that the repository would need to ensure isolation of the wastes for several hundred years only, rather than for several hundreds of thousands of years.

II.7.4 The Energy Amplifier

The Energy Amplifier is the brainchild of the 1984 Nobel prizewinner Carlo Rubbia of CERN. His proposal is for an accelerator-driven power reactor utilizing abundant thorium as the primary nuclear fuel. While similar in technical scope to the use of ADS technology for the fast neutron transmutation of nuclear wastes, the energy amplifier concept extends that ambition beyond merely waste disposal to include a radical and innovative approach to future nuclear power production. Figure II.7.10 is a schematic of the outline design for the nuclear parts of the energy amplifier. The

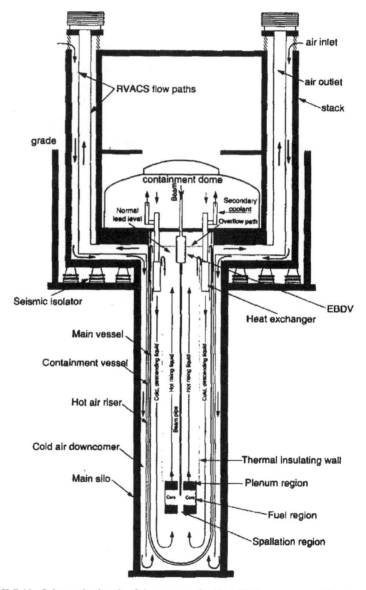

Figure II.7.10. Schematic sketch of the proposed 1500 MWth energy amplifier. Spallation neutrons are produced at the bottom of the 25 m tall device (source: J-P Revol [93]).

similarity to the French CEA waste transmutation facility illustrated in figure II.7.9 is immediately clear.

While the energy amplifier has technological similarities to ADS nuclear waste burners, it is the extension of the ADS approach to focus on the production of economic electrical power that is key to the energy amplifier concept. The natural comparison therefore is between the energy amplifier and conventional uranium fuelled thermal critical reactors. The following aspects of that comparison are particularly noteworthy.

- As a sub-critical accelerator-driven system the energy amplifier proponents point to its greater intrinsic safety and greater ease of control. In the event that the reactor becomes unstable, then all that is required is that the operators turn off the accelerator driving the machine. Two caveats are, however, worth emphasizing here. The first is that as the history of reactor accidents in critical systems shows a key concern in any reactor system is the management of decay heat once the nuclear fission process itself has stopped [62]. The second is that the possibility of nuclear criticality is determined not only by the mass of fissile material, but also by its geometry: the amount of material that might be critical when in a spherical shape would be unlikely to be critical if stretched into a long rod or flattened like a pancake. Designers of the energy amplifier seem confident that the system does indeed have superior safety attributes when compared with conventional reactor systems. It is important to remember, however, that a sub-critical assembly is not inevitably a safer technology.

- A second key difference from conventional uranium-fuelled thermal critical reactors is that the energy amplifier uses thorium rather than uranium fuel. Thorium is roughly three times more abundant than uranium and as such is posited as the basis of a wholly new nuclear fuel cycle in the event that economic uranium resources are depleted. The use of thorium as a fission reactor fuel is interesting because thorium is not fissile. The key to the use of thorium is neutron capture by thorium-232 to form fissile uranium-233. While it is possible to construct a critical thorium/plutonium reactor the safety margins would be much tighter than in conventional uranium reactors as the fraction of delayed neutrons is much lower in the thorium-based concept. In an accelerator-driven system, where the level of criticality is maintained by an external neutron source, such concerns are much less burdensome [93]. Thorium has the added benefit of generating very few higher actinides (Po, Am, Cm etc.) when used as a nuclear fuel.

- The energy amplifier makes use of fast neutrons in order that it may also be applied to nuclear waste incineration. High rates of fuel burn-up will be achieved and as the ratio of fission to neutron capture is larger for fast neutrons, fewer secondary wastes will be created [93].

II.7.5   Thermal neutron waste management

In section II.5.3 we discussed the fuel cycle benefits of the 'DUPIC' fuel cycle that can be used by the AECL Advanced CANDU Reactor (ACR). This allows for high burn-up of re-fabricated used light water reactor fuels. The ACR is also well suited to the high burn-up use of mixed oxide fuels, eliminating some of the difficulties associated with MOX in LWR systems. While thermal reactors will never be able to match the incineration attributes of fast neutron systems, it is clear that improvements over more conventional thermal fuel cycles can be achieved.

One thermal system with particular benefits in fuel cycle management is the General Atomics 'Deep Burn' concept in a high-temperature gas-cooled reactor [94]. These ideas are closely related to the gas turbine–modular helium reactor discussed in chapter 6. Related technologies are also discussed in chapter 8.

II.7.6   Accelerator-driven systems, transmutation and new build

By 'new build' we mean the construction in western Europe and North America of new nuclear reactors for commercial electricity production. In this chapter we have emphasized the importance of accelerator-driven systems (ADS) in the context of nuclear waste transmutation. Despite the numerous research challenges and the significant road to be travelled before these systems are realized, they are more immediate and seemingly achievable than the grander ambitions for radical ADS energy systems, such as the energy amplifier. While the technological links between the energy amplifier and accelerator-driven waste transmutation facilities are clear, such synergies are certainly in the distant future. More immediately one is drawn to ask what are the links between accelerator transmutation of wastes (ATW) and conventional new nuclear build. As discussed in Part I, it has become almost axiomatic to discuss nuclear waste management via the lens of the legacy waste problem. By arguing that waste management should be unrelated to future nuclear power production and to any decisions on new build, it is hoped to engage those opposed to nuclear power and concerned about the environment in the current nuclear waste debate. While some may regard this as politically expedient it is not necessarily the technically optimal way in which to frame the debate, especially for ATW-based approaches. As French experts first brought to my attention, in the absence of nuclear new build, an ATW approach would be expensive, slow and difficult. By implication if there were to be an expanding nuclear industry with new build then ATW approaches would be correspondingly cheaper, quicker and easier to deploy. Key to these ideas is that with an ongoing and substantial nuclear industry it is easier to find the staff, the capital resources and to develop the physical infrastructure and overall capacity for complex endeavours of this type.

Recognizing these synergies, one US environmental organization has described ATW as the 'Trojan Horse' of the nuclear industry [95]. This rich metaphor intends to imply that transmutation technology, sold on its ability to tackle legacy wastes of concern to us all, is in fact a technology whose real underlying purpose is to propel us into a renaissance for nuclear power. While synergies would undoubtedly exist between accelerator-driven systems and a nuclear renaissance, in this author's opinion it would be incorrect to imply that the proponents of ATW are deliberately setting out to deceive. This author regards the synergies to be an important opportunity rather than any kind of a Trojan Horse.

ATW development would benefit from a new nuclear build programme, but does not require it. Similarly new nuclear build may in the very long term be favoured by ATW research, but the role of ATW in any new nuclear build decisions at this time must be very minor indeed. It would be wrong to deny that there are links between ATW and new nuclear build, but such links as there are, are weak, and certainly not defining for future policy.

Chapter 8

Generation IV

In the first seven chapters we have considered technologies and policies that would be central to any moves towards a nuclear renaissance in Europe and North America. In doing so we have emphasized near-term technologies with relatively well understood engineering issues and economics and have been considering technologies from the second and third generations of nuclear power. It has become standard to describe the successive developments in nuclear power plant engineering in terms of the following generations of technology.

- Generation I: early prototype reactors of the 1950s and early 1960s (e.g. Shippingport and Magnox).
- Generation II: commercial power reactors of the 1970s and 1980s (e.g. LWR of both PWR and BWR types, CANDU and AGR).
- Generation III: advanced light water reactors (e.g. systems maturing in the late 1990s such as ABWR, AP1000, ACR and EPR).
- Generation III+: these are innovative new reactor designs showing promise for the near term future (e.g. PBMR and GT-MHR high-temperature reactors).

It is expected that Generation III+ deployments will come to an end by approximately 2030, to be superseded by a next generation of economic, safe, low waste and proliferation-resistant plants to be known as 'Generation IV'.

For some, Generation IV technologies represent the core of any nuclear renaissance. They would argue that the issues of nuclear renaissance are synonymous with the issues surrounding Generation IV and that the success-ful deployment of Generation IV technologies will be the defining stage of any nuclear renaissance. This author, however, disagrees. I believe that the issues of near-term deployment of Generation III technologies are the key determinants of the future of commercial nuclear power. Without plans for new build starting in this first decade of the century, it is hard to imagine that nuclear power in the west has any future whatsoever. It is unlikely that North America and Europe would be able to deploy Generation IV technologies in the years after 2020 if nuclear power had not already experienced substantial new build based upon Generation III and Generation III+ technologies. In part this is a consequence of the continued erosion of the skills base and the concomitant loss of industrial and regulatory capacity that will occur without near-term new build.

It is important to acknowledge that those working on Generation IV technologies for deployment in 2030 are, of course, extremely conscious of today's prerequisites in terms of nuclear skills and underpinning research. That is to say, while Generation IV looks to the long term, all acknowledge that it raises technology and policy issues for the present.

Some might argue that the maintenance of nuclear industrial capacity, in the spirit of keeping the nuclear option open, could allow for a late renaissance based upon generation III+ and Generation IV technologies in the 2020s and 2030s without intervening new build. However, it seems more likely that defining decisions towards or away from a nuclear renaissance are required in the nearer term. Such decisions will concern new build using existing Generation III technologies.

In the past few years, the term Generation IV has assumed a second, much more specific, meaning than that referred to above. It has come to be associated with a particular international collaborative group led by the United States Department of Energy. This group, known formally as the Generation IV International Forum (GIF), was chartered in the summer of 2001 following an earlier series of preparatory meetings starting in January 2000. Initially led by an inner policy group, the GIF brings together governmental representatives from ten countries (Argentina, Brazil, Canada, France, Japan, South Korea, South Africa, Switzerland (a late entrant), the United Kingdom and the USA).

In December 2002 the GIF published a major report under the title *A Technology Roadmap for Generation IV Nuclear Energy Systems* [96]. In March 2003 the Roadmap was presented to the US Congress. The Roadmap summarizes the work of the GIF up to the end of 2002 and presents an outline path towards eventual deployment of Generation IV systems based upon international collaboration. The US leadership of the GIF project shaped the process by which competing technologies were assessed for Generation IV development. Furthermore, the US leadership of the GIF process is clear from the scope and emphasis of the GIF documentation.

Prior to the launch of the GIF, the US DoE had set up a Nuclear Energy Research Advisory Committee (NERAC). A NERAC study published in October 2001 had focused on technologies deployable by 2010 [97]. Eight candidate designs were assessed in that process (ABWR, SWR1000, ESBWR, AP600, AP1000, IRIS, PBMR and GT-MHR) against six criteria: regulatory acceptance, industrial infrastructure, the commercialization plan,

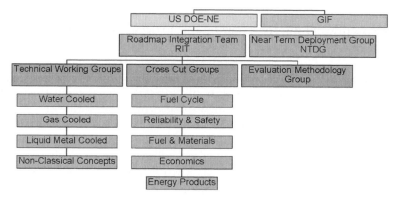

Figure II.8.1. Structure of the US Department of Energy led Generation IV project [98].

the cost-sharing plan, economic competitiveness, and fuel cycle industrial structure. In this book we do not attempt to survey all possible nuclear power designs that might contribute to a nuclear renaissance. We merely seek to explore the range of technology and policy issues affecting new build decisions. Consequently this book does not emphasize the small Westinghouse PWR (100–300 MWe) PWR known as IRIS, the ABB Westinghouse advanced PWR System 80+, or the GE ESBWR. This latter design has many similarities to the ABWR design discussed in chapter 5.

The NERAC work informed the early meetings of the GIF, but it was immediately acknowledged that there was a need to broaden the range of designs under consideration to reflect international interests. The GIF therefore created the International Near Term Deployment Group. This extended the NERAC list of near-term technologies to 16 designs including both the Canadian ACR-700 and the EPR.

The enlarged GIF international near-term deployment list defines a set of reactors from which any nuclear renaissance based upon generation III and generation III+ technology would need to emerge.

However, as its name suggests, the primary attention of the GIF is not devoted to near-term Generation III and Generation III+ issues, but rather is devoted to the longer-term task of advancing true Generation IV technologies. Contenders for Generation IV status have been assessed by the GIF against eight equally weighted goals [96]. See text box *Technology goals*.

The GIF itself acknowledges that the sustainability criteria adopted are narrow and do not span the range of considerations currently accepted internationally as being central to balanced sustainable development [96]. The GIF accepts the Brundtland formulation of sustainable development, namely that our actions should meet the needs of the present while not compromising the ability of future generations to meet their own needs [99]. However, the Brundtland report of 1987 built the fundamental

---

Technology goals for Generation IV nuclear energy systems [96]

Sustainability—1. Generation IV nuclear energy systems will provide sustainable energy generation that meets clean air objectives and promotes long-term availability of systems and effective fuel utilization for worldwide energy production.
Sustainability—2. Generation IV nuclear energy systems will minimize and manage their nuclear waste and notably reduce the long-term stewardship burden in the future, thereby improving protection of the public health and the environment.

Safety and reliability—1. Generation IV nuclear energy systems operations will excel in safety and reliability.
Safety and reliability—2. Generation IV nuclear energy systems will have a very low likelihood and degree of reactor core damage.
Safety and reliability—3. Generation IV nuclear energy systems will eliminate the need for offsite emergency response.

Economics—1. Generation IV nuclear energy systems will have a clear life-cycle cost advantage over other energy sources.
Economics—2. Generation IV nuclear energy systems will have a level of financial risk comparable with other energy projects.

Proliferation resistance and physical protection—1. Generation IV nuclear energy systems will increase the assurance that they are a very unattractive, and a least desirable, route for diversion, or theft, of weapons-usable materials. Furthermore they will provide increased physical protection against acts of terrorism.

---

framework for sustainable development based upon three equal criteria—the economic, the environmental and the social. These three pillars of sustainable development are sometimes referred to as 'the three Ps': Prosperity, Planet and People. The GIF sustainability criteria focus exclusively on environmental sustainability with considerations for economic sustainability arguably covered under other goals. The economic goals include reference to the full life cycle of the technology including long-term waste management. However, the pillar of sustainability that is completely omitted in the GIF goals is the concept of social sustainability. In this author's opinion this omission is the most serious weakness in the GIF approach, an approach that in other respects has been a model of ambition and vision. It seems that the lack of explicit consideration of the social and political needs and concerns of current and future generations has already started to shape the balance of work undertaken by the GIF and the projects favoured by it. The GIF process has neglected to factor in what types of nuclear power plants or fuel cycles the various national publics and their representatives might actually want in the years after 2030. Experiences with consensus conferences and other forms of public engagement have shown that ordinary citizens are willing to participate in, and to contribute to, policy making and prioritization in these intensely technical areas. For instance, any strong public aversion to separated plutonium in the fuel cycle or to nuclear waste 'disposal' (as opposed to monitored and retrievable storage) would be important factors to incorporate into Generation IV planning. At present such inputs seem to be lacking. Generation IV priorities, having been developed in the tradition of engineering-led design and shaped by technocratic policy-making, risk falling at the final hurdle. The risk is real that the publics of liberal western democracies might refuse to allow Generation IV new build for reasons which are deeply held, widespread and visible, but which the technocrats have dismissed as idiosyncratic and naive.

Technocracy is defined as 'rule or control by technical experts' [100]. The GIF approach seems to be a technocracy tempered by national policy priorities rather than political forces. It is noteworthy that the GIF goals make no explicit reference to what is arguably the greatest factor driving the need for Generation IV technologies—fossil fuel-based CO_2 emissions and their impact on global climate change. Rather, the first GIF sustainability goal emphasizes the role of Generation IV technologies in delivering 'cleaner air'. It is important at this point to note that this clause, in a US context at least, does not include consideration of CO_2 emissions. The US Environmental Protection Agency regulates emissions of numerous pollutants and harmful gases, but notably CO_2 is not among them. The GIF goals seem to be shaped by current US political considerations that would seem to pull away from best science and engineering practice regarding both global climate and the dangers of uncontrolled greenhouse gas emissions.

In considering the GIF goals we have so far emphasized the omissions and compromises in the project. The real strengths of the activity, however, deserve mention. The GIF Roadmap is a notably ambitious document that describes the path ahead for six technologies that have passed assessment against the eight GIF goals. These six technologies have emerged from a process that considered 30 water-cooled concepts, 17 gas-cooled concepts, 33 liquid metal-cooled concepts and 17 'non-classical' concepts [98]. The six Generation IV systems that survived GIF scrutiny indicate no sign of defensiveness or defeatism on the part of the nuclear industry. In fact the GIF priority list reveals self-confidence reminiscent of the hubris of the nuclear barons in the 1960s and 1970s. A Uranium Information Centre briefing paper nicely summarizes the selected GIF technologies: most employ a closed fuel cycle on the grounds that this maximizes the resource base and minimizes the amount of high-level wastes needing to be sent to a repository [101]. The GIF proposes that Generation IV will probably require an extension of reprocessing with new techniques being industrialized. This contrasts starkly with British media speculation concerning the fade out and decline of reprocessing [102]. Also, the enthusiasm of the GIF for closed fuel cycles contrasts markedly with the findings of the MIT *Future of Nuclear Power* study group, which rejects any extension of nuclear fuel reprocessing [103]. The long-standing US policy aversion to reprocessing in the nuclear fuel cycle is visible, within the roadmap document of the GIF in the tendency to refer to 'actinide management', rather than 'next generation reprocessing' as such issues might be termed in Europe.

Even more amazing than the interest in closed fuel cycles is the fact that three of the six reactors being proposed by the GIF are fast reactors [101]. Some have the possibility of being developed as fast, epithermal or thermal reactors. Only two are clearly designed to operate as thermal reactors similar to all western commercial nuclear power plants in operation today. The ambition and vision of the GIF does not stop there, because only one of the six proposals is cooled by light water. Two are helium-cooled and the other three have metal or molten salt cooling. The GIF recommendations for Generation IV are indeed radical and ambitious, albeit in ways that refer back to several of the dreams of the nuclear technologists in the years before the slowing of western nuclear power.

The six technologies selected for research and possible development by the GIF as 'most promising systems' are as follows [96].

II.8.1   The gas-cooled fast reactor (GFR)

This concept builds upon the gas turbine–modular helium reactor discussed in chapter 6. That design, however, uses a thermal neutron spectrum and a once-through fuel cycle. The GIF recommends building upon Generation III+ experience with high-temperature gas-cooled reactors, such as the GT-MHR and the PBMR, and to extend the concept to incorporate the use of a fast neutron spectrum and a closed fuel cycle.

The use of a fast neutron spectrum allows the use of fertile fuels as well as fissile fuels. The successful use of depleted uranium as a fuel would greatly increase the nuclear fuel resources available and so enhance the long-term sustainability of nuclear power. This together with a move to a closed reprocessing-based fuel cycle would allow for a very significant reduction in the total quantities of high-level radioactive wastes requiring management or disposal [96]. The GIF estimates that the conversion of mined reserves into useful heat will be two orders of magnitude more efficient than with thermal spectrum gas reactors with once-through fuel cycles. Moving towards a closed fuel cycle in high-temperature gas-cooled reactors also requires some thought as to the form of nuclear fuel to be used. At present both

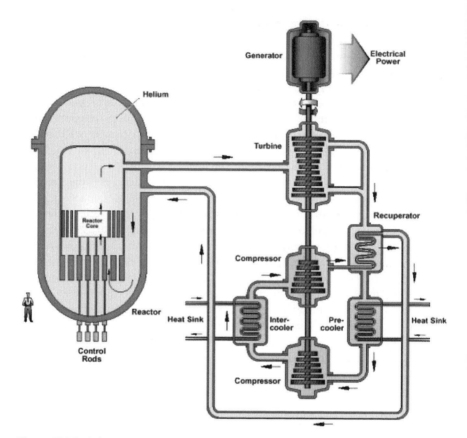

Figure II.8.2. Schematic of the GIF gas-cooled fast reactor (source: Idaho National Laboratory).

GT-MHR and PBMR intend to make use of Triso granular fuel formed into respectively prismatic blocks or 'pebbles' (see chapter 6). Triso fuel is, however, poorly suited to reprocessing given the in-built silicon carbide barrier layer in the fuel granules. These concerns have prompted the GIF to adopt a more open-minded approach to possible fuel forms. Currently ceramic fuels and ceramic-clad fuel kernels made from actinide compounds are under consideration complementing developments in Triso fuel [104]. Significant research will be required to establish novel fuels suitable for use at very high temperatures in a fast neutron flux.

As figure II.8.2 makes clear, the GFR concept builds upon the elegant General Atomics proposal employed in the GT-MHR for a single combined turbine and compressor shaft.

The GIF predicts that a conceptual design for an entire GFR prototype system can be developed by 2019 with an international prototype in operation by 2025 [96].

II.8.2 Very-high-temperature reactor system (VHTR)

The other reactor system that builds upon Generation III+ experience with high-temperature gas-cooled reactors is the very-high-temperature reactor system (VHTR) (see figure II.8.3).

This proposal suggests a graphite-moderated, helium-cooled reactor with a once-through uranium fuel cycle [104]. Attention is also being given to the use of mixed uranium/plutonium fuel cycles. The VHTR is described as a next step in the evolutionary development of high-temperature gas-cooled reactors [96]. The GIF Roadmap acknowledges the technical debt of the VHTR to the Japanese prismatic HTTR and the Chinese pebble bed HTR-10 as well as earlier HTGR experiments. It acknowledges the current development of the PBMR and GT-MHR, but does not emphasize the fact that these two existing projects clearly fit the GIF description for the VHTR. Central to the research needs of the GIF VHTR are the properties of Triso fuel, a topic currently widely studied independent of the existence of the GIF.

The GIF adds one significant new idea to HTGR thinking in the context of the VHTR. The new idea is sufficiently imaginative that it alone takes the VHTR concept from the arena of Generation III+ firmly into the zone of Generation IV. The idea is that the VHTR could separate commercial nuclear energy from its technological bedfellow, the electricity industry.

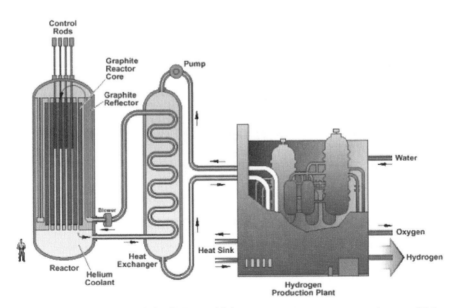

Figure II.8.3. Schematic of the GIF very-high-temperature reactor system (source: Idaho National Laboratory).

Thus far this book has focused almost exclusively on the idea that the form and impact of a nuclear renaissance will relate inevitably to the economic and policy issues surrounding electricity generation. The GIF VHTR concept frees nuclear energy from these difficulties and allows nuclear energy directly to address a different societal need that will probably grow to prominence on a timescale similar to the timescale of Generation IV deployment, the resource depletion of fossil fuels needed for transport. The GIF Roadmap discusses the possibility that the VHTR might be used to create hydrogen to fuel a hydrogen economy without the need for wasteful energy conversion steps involving electricity (see section I.3.4).

II.8.3   Supercritical water-cooled reactor (SCWR)

The GIF acknowledges that the supercritical water-cooled reactor (SCWR) (figure II.8.4) builds upon much experienced from the Japanese supercritical

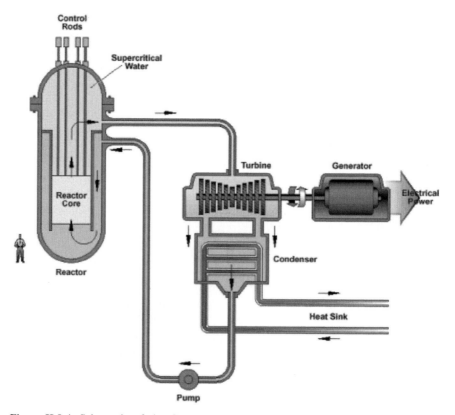

Figure II.8.4. Schematic of the GIF supercritical water-cooled reactor (source: Idaho National Laboratory).

light water reactor (SCLWR) [96]. Key to the SCWR concept is that the water used to cool the reactor is above its thermodynamic critical point of 374 °C and 22.1 MPa. At such high temperatures and pressures water no longer exists as two phases (liquid and steam); rather it comprises a single compressible fluid of smoothly varying density in response to changes in pressure or temperature.

The SCWR benefits from much pre-existing technological development. It benefits from experience gained with LWR design and operation (particularly in the development of BWRs). It also benefits from the development of supercritical water energy extraction in fossil-fuelled electricity generation. SCWR systems offer the prospect of significantly higher thermodynamic efficiencies than are available from conventional LWRs. The GIF Roadmap estimates that the SCWR will achieve thermal efficiencies of 44% compared with 33–35% for conventional LWRs. Supercritical water can absorb more heat than steam per unit mass or volume. That is, its enthalpy is higher. This allows for a reduction in the coolant mass flow required and therefore a reduction in the size of the reactor coolant pumps, the piping and the containment building housing the key equipment. As the process of energy extraction deals with a single fluid phase there is no need for expensive steam dryers, steam separators, recirculation pumps and steam generators [96]. The far greater simplicity of the SCWR results in far lower capital costs. The GIF estimates that an SCWR could be constructed for approximately US $900 per kWe, that is for roughly half the cost of Generation III LWRs [96].

Design and licensing of an industrial scale SCWR is likely to be far from straightforward. There remain significant unknowns and challenges in the area of materials for nuclear components in a supercritical water environment. While the SCWR builds upon a large body of experience in the design and operation of LWRs, it pushes LWR technology, placing far larger physical stresses on the plant under normal operations. The SCWR operating pressure of 25 MPa, the order of magnitude changes in coolant density and the high operating temperatures at around 500 °C all require investigation and engineering development, particularly in the area of fuel claddings and structural materials, to ensure safe and reliable operations.

The GIF reports that, if all goes well, a conceptual design should be available by approximately by 2015 or 2016 [96].

While initially the GIF proposes a thermal SCWR, attention is also being focused on the possibility of a fast neutron variant of the SCWR. This would be similar to the thermal SCWR, but would have no structural moderator complementing the liquid water flowing into the pressure vessel. Moving to a fast neutron spectrum would bring with it the usual advantages of lower fuel resource depletion, but at the price of far greater radiation damage to the structural elements of the reactor. The Roadmap document estimates that the doses to the cladding and structural materials could be larger than in the thermal version by a factor of five or more.

II.8.4   Sodium-cooled fast reactor (SFR)

The concept of a sodium-cooled fast reactor (figure II.8.5) is the most mature of all the Generation IV technologies proposed in the GIF Roadmap. Sodium-cooled fast reactor demonstrators have ranged from the US 1.1 MWth EBR-1 of the 1950s to the 1200 MWe Super Phenix reactor in France in the 1980s [96]. At various times France, Japan, Germany, the UK, the USA and Russia have pursued such programmes. The world has more than 300 reactor years of fast neutron reactor operation experience already. At present, however, only Japan and Russia have substantial ongoing plans of their own. Most of the countries listed abandoned their efforts in the face of falling uranium prices and the high levels of technological challenge involved in developing the technology. It seems likely that SFR technologies will not become economic until such time as uranium prices rise significantly. In the meantime technology innovations should be dedicated to reducing capital costs of construction, improving intrinsic safety (particularly via a greater use of passive systems), and further work on the fuel reprocessing system essential for the sustainable operation of fast reactor systems. There is much interest within GIF in moving the SFR fuel cycle towards pyrochemical reprocessing and away from aqueous approaches. This has the benefit of improving the proliferation resistance

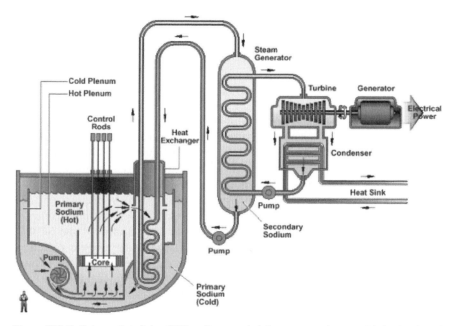

Figure II.8.5. Schematic of the GIF sodium-cooled fast reactor (source: Idaho National Laboratory).

of the technology. A developed SFR would be able to play an important role in the management of currently surplus materials arising from Generation II and III technologies. For example, SFR technology would have much to offer the UK in terms of disposal (and use) of the UK's legacy of separated civil plutonium. For countries with legacies arising from the operation of a once-through fuel cycle the SFR would be able to allow for the unlocking of unused energy remaining in the spent fuel. The SFR was originally strongly supported by France and the UK as a means by which to close their reprocessing-based fuel cycles. Indeed the prospect of eventual SFR power generation was a major driver behind the development of PUREX-based reprocessing in these countries. The fuels needed by the SFR include the plutonium mentioned above, but also uranium-238, which has no beneficial use in conventional thermal neutron fuel cycles.

The proposed SFR would operate at approximately 550 °C. Importantly the reactor vessel would be held at a pressure only slightly above atmospheric pressure, the over-pressure required being merely that needed to allow the circulation of the liquid metal coolant. As a power station, the SFR technology could be deployed in either of two sizes: a medium sized machine in the range 150–500 MWe, and a large-scale machine in the range 500–1500 MWe.

II.8.5   Lead-cooled fast reactor (LFR)

The lead-cooled fast reactor (figure II.8.6) borrows much from the Russian concept of the BREST fast reactor discussed in chapter 7. That technology in turn benefits from much experience gained by the Soviet Navy in the use of lead–bismuth eutectic cooling in naval propulsion reactors in its Alpha class submarines [96]. Russia is not a formal partner in the GIF although links between the Russians and the American-led GIF process do exist. Expertise with lead-cooled and lead–bismuth-cooled reactor systems among GIF member countries is restricted to the US STAR and the Japanese LSPR programmes respectively [101]. In addition, much of the work on materials undertaken in western countries in connection with sodium-cooled fast reactor technology can most probably be applied to the LFR project.

As a fast-spectrum reactor with full actinide fuel recycle the system has much in common with the sub-critical fast-spectrum nuclear waste transmutation technologies discussed in chapter 7. The GIF restricts itself to consideration of critical fission-based systems, and hence does not include technologies such as the energy amplifier. However, it is through the LFR proposal that the greatest synergies between Generation IV and the energy amplifier will be possible.

The LFR reactor outlet temperature is likely to range between 550 and 880 °C, depending upon the use of advanced materials [104]. If the higher

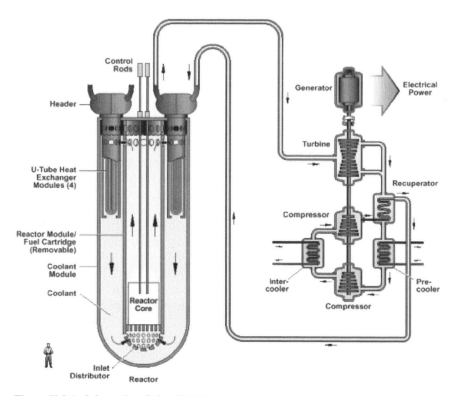

Figure II.8.6. Schematic of the GIF lead-cooled fast reactor (source: Idaho National Laboratory).

temperatures can be attained (in this version the coolant would not include bismuth), the LFR will be well suited to the thermochemical production of hydrogen. It is envisaged that LFR technology could proceed via step-wise phased development with initial applications being dedicated to electricity production and with later development focusing on hydrogen production [96].

In the short term it is proposed that the LFR would use metal alloy fuel or nitride fuel, if available. Possible metal alloy fuels and the back-end recycling of uranium, zirconium and (using the US definition) TRU waste would also benefit from likely synergies with developments occurring in the parallel SFR programme [96].

Nitride fuels are required for the high temperature version of the LFR. Moves to full actinide recycle in nitride fuels will require significant new research and development. In this context the GIF Roadmap recommends moves towards non-aqueous pyrochemical reprocessing.

Among the six GIF concepts, the LFR is distinguished by the possibility that it can be developed in three different sizes: a large monolithic plant rated at 1200 MWe, a small to medium sized modular system at 300–400 MWe, and most interestingly of all, a very small system rated at 50–150 MWe with very long refuelling intervals [96]. This last system, known as the LFR 'battery', would utilize modular factory-built replaceable reactor cores. The fuel cores needing to be changed only every 15–20 years. The only refuelling would be that associated with the changeover of an entire reactor core module or 'battery'. In some applications one could even imagine LFR technologies being deployed with no intention to ever refuel the reactor. The small, modular battery concept is extremely well-suited to deployment in developing countries with no interest in developing their own nuclear fuel cycle infrastructure. It would, for instance, be a technology well suited to delivering the obligations of nuclear weapons states under the Nuclear Non-Proliferation Treaty to share peaceful nuclear technologies with developing countries that are signatories to the NPT and who have constrained their own nuclear ambitions.

II.8.6   Molten salt reactor (MSR)

The molten salt reactor (MSR) (figure II.8.7) is the most technologically radical of all the six GIF preferred technologies. At the heart of the radical thinking is the fact that the nuclear fuel is not in the form of prefabricated engineered fuel assemblies; nor even is the fuel solid. Rather, the fuel for

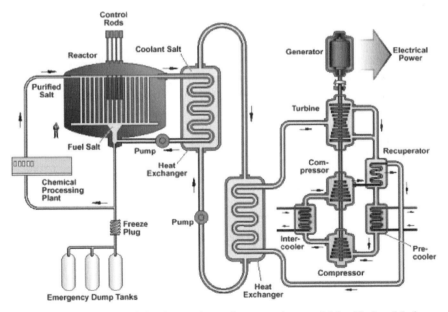

Figure II.8.7 Schematic of the GIF molten salt reactor (source: Idaho National Laboratory).

the MSR takes the form of a liquid pumped through the reactor core. The liquid molten fuel is composed of uranium or plutonium fluorides dissolved in sodium or zirconium fluorides, selected for optimal neutronic properties.

The GIF MSR would be characterized by the following attributes [96]: it would be a large-scale (1000 MWe) base load generator; it would utilize graphite moderated epithermal neutrons; it would operate at high temperature (700–850 °C), permitting hydrogen production; it could operate flexibly with a range of fuel compositions. In particular, the reactor could benefit from fertile isotopes in the fuel mix, such as uranium-238 or thorium-232. Also the MSR is well suited to actinide disposal, although this would require offsite fuel processing. Otherwise it would be expected that on-site fuel processing would be sufficient to maintain healthy neutronics of the salt such that it could be used for several years without being changed. The MSR does not produce significant actinide wastes owing to very high burn-up. This offers good proliferation resistance and excellent primary fuel resource efficiency [101].

It is believed that, despite its radical and somewhat experimental conceptual status, the MSR concept offers good prospects for reactor safety [96]. The reactor generates almost negligible vapours and so can be regarded as a single liquid phase at all points. Over long periods the build up of radioactive gases could be cause for some concern, but the volumes of gas should never be sufficient to jeopardize the fundamental design concept of a uniform liquid fuel. In the event of a serious problem, a 'freeze plug' at the base of the reactor would melt passively allowing the reactor fuel to drain safely to a set of emergency dump tanks (see figure II.8.7).

The MSR is likely to employ two separate molten salt circuits. The primary circuit would contain the nuclear fuel and resulting fission products, while a secondary circuit would exchange heat with the primary circuit and then pass that heat to a second heat exchanger. The heat exchanger could pass the heat to a gas coolant that would pass through a Brayton cycle gas turbine or alternatively pass it to a system designed to catalytically generate hydrogen in a high temperature reaction. The use of primary and secondary molten salt circuits is believed to offer improved safety and easier decommissioning and maintenance.

Although a series of research molten salt reactor did operate successfully in the United States in the 1950s and 1960s. These reactors were originally developed with the intention that they might one day be used to power aircraft, hence the name of the first such reactor: the Aircraft Reactor Experiment. At the end of the 1990s the second such American reactor, the 8 MWth Molten Salt Reactor Experiment, proved a concern for decommissioning after it had stood idle for almost 30 years. It required attention to remove the radioactive salts that had lain undisturbed in tanks close to the reactor throughout that time. The challenges for safely decommissioning this aged facility were substantial. Despite the low vapour pressure in molten salt reactors over the long period there had been significant build-up of radio-active uranium hexafluoride gas and hazardous fluorine. In addition, special care was required to avoid criticality incidents in defuelling the facility [105].

These difficulties relate, in part, to a large set of unknowns regarding the fundamental design and operation of a new MSR concept. Much remains to be understood about the long-term fuel chemistry, the materials behaviour of reactor components, risks of metal clustering and precipitation, and the necessary salt processing, both on-site and offsite.

II.8.7   Generation IV and the hydrogen economy

The potential for using nuclear energy for hydrogen production has been discussed in chapter 3. Generation IV International Forum strongly endorses this vision. Of the six preferred technologies in the GIF Roadmap four are deemed suitable for thermo-chemical hydrogen production. Three of these four have expected maximum operating temperatures around 800 °C. Maximum operating temperatures of around 800 °C are problematically low, given that the efficiency of thermo-chemical hydrogen production increases rapidly with increasing temperature. These three concepts (GFR, LFR and MSR) are primarily planned to be electricity generating technologies.

The truly exciting technology for hydrogen production is the very-high-temperature reactor. With expected operating temperatures in the range of 1000 °C, this technology has the potential to be led by the needs and opportunities of the hydrogen economy. The GIF estimates that the VHTR could reach the stage of a conceptual design by 2015. This timing is well suited to the probable needs of those pursuing hydrogen as an energy vector for vehicular transport. As discussed in chapter 3, if the links binding nuclear energy and the electricity industry can be weakened, then it seems likely that a wider range of more ambitious new nuclear energy technologies can be developed. Generation IV's emphasis on hydrogen is an important positive step in that direction.

II.8.8   Generation IV and distributed generation

As discussed in the preceding section the GIF allows for an important break between commercial nuclear energy and the production of electricity. It is all the more interesting therefore that the GIF does not focus its attention on the idea that the future of nuclear electricity generation is distributed and small-scale in the manner prophesied by Walt Patterson in his book *Transforming Electricity* [106]. Of the six preferred technologies in the GIF roadmap two are clearly from the traditional large monolithic paradigm of commercial nuclear power (MSR and SCWR) and two are clearly mid-scale (GFR and

VHTR). One technology (SFR) has the potential to be either large-scale or mid-scale and only one has the potential to be truly small-scale and modular (LFR), although that technology could also be constructed in a traditional monolithic large-scale configuration [98].

II.8.9   Technologies beyond Generation IV

It is noteworthy that neither accelerator-driven systems (chapter 7) nor nuclear fusion energy (part 3) are recommended by the Generation IV International Forum in its Roadmap [96]. This is surely a consequence of time-scales, as neither technology is likely to have a workable power station concept by 2025. These technologies are frankly closer to Generation V than Generation IV. It would, however, be most unfortunate if the lack of emphasis in the GIF process for these technologies were actually to diminish resources and effort devoted to their development. As it is, it seems that both these long-term technologies have sufficient momentum and resources to make good progress even in the absence of support from Generation IV. The momentum of fusion research is maintained by the international ITER programme. The momentum for accelerator-driven systems and the energy amplifier is not as strong and it is this project in particular that needs to be given attention to ensure that it does not suffer as a consequence of not being a GIF priority.

II.8.10   The UK and Generation IV

In early 2004 details of UK participation in Generation IV research are still unclear and under discussion. The Department of Trade and Industry, BNFL and NNC have together identified that the UK should concentrate on three of the six preferred GIF technologies. The three technologies well suited to UK involvement are [98] as follows.

- The very-high-temperature reactor (VHTR), given UK involvement in the Pebble Bed Modular Reactor, previous HTGR experience at Winfrith and the long-standing experience with gas-cooled reactors.
- The gas-cooled fast reactor (GFR), which builds upon the extensive experience of the UK with gas-cooled reactors, but also on UK research experience with fast reactors at Dounreay in Scotland and elsewhere (e.g. ZEBRA at Winfrith).
- The sodium-cooled fast reactor (SFR), which builds upon UK fast reactor experience, much of which employed liquid sodium coolant.

In its first few years the GIF process has maintained an impressive momentum and hit its project milestones on schedule. It is therefore noteworthy that the GIF process requires that actual research related to the six preferred technologies should start in April 2004. On page 79 of the Roadmap document the GIF makes clear its assessment that research into each of the six preferred technologies will require roughly US $1 billion. In April 2004 the UK is still to determine its cash and intellectual property contribution to the GIF research programme. Noting that the rough estimate of US $1 billion per project and the likely 20-year duration of GIF research it seems likely that the UK will need to find sums of the order of US $10 million per annum in order to buy a proper seat at the table. As such, Generation IV would play a significant part in UK government policy to keep the nuclear option open.

The importance of the VHTR concept for developments in nuclear hydrogen production is of potentially great benefit to the UK. It is pleasing that the UK is taking an active interest in the Generation IV VHTR project. If nuclear hydrogen production is eventually demonstrated to be industrially and economically viable then the UK, with its strong capacities and capabilities in the oil and petrochemical industries, will be very well placed to develop nuclear hydrogen for widespread vehicle use.

Part II references

[1] Argonne National Laboratory, Engineering Research History (as of June 2004: www.era.anl.gov/history/pwr.html)

[2] Paulson C K 2002 'AP1000: set to compete', *Nuclear Engineering International* October

[3] Cummins W E and Mayson R 2002 *Westinghouse AP1000 Advanced Passive Plant: Design Features and Benefits*, paper presented at the European Nuclear Congress, Lille, France

[4] Nuclear Energy Agency, Committee on the Safety of Nuclear Installations 2002 *The Use and Development of Probabilistic Safety Assessment in NEA Member Countries*, NEA/CSNI/R(2002)18, pp 176–177, Paris (as of August 2004: http://www.nea.fr/html/nsd/reports/csnirepindex.html)

[5] British Energy, Press Release, 2 November 2001

[6] Atomic Energy of Canada Ltd, *Economic Electricity for the 21st Century*, Marcom Catalogue #1-21, August 2001

[7] AECL, private communication

[8] Hewitt G F and Collier J G 2000 *Introduction to Nuclear Power* (London: Taylor and Francis) p 77

[9] Lamarsh J R and Barratta A J 2001 *Introduction to Nuclear Engineering* (Upper Saddle River, NJ: Prentice Hall) pp 163–168

[10] British Energy, *Replace Nuclear With Nuclear*, submission to the Performance and Innovation Unit's Energy Review, September 2001 (as of June 2004: http://www.british-energy.com/corporate/energy_review/energy_submission120901.pdf)

[11] Atomic Energy of Canada Ltd, *The ACR—The ideal solution for plutonium dispositioning*, AECL Business Development Office, briefing sheet

[12] Bruce Power, Press Release, *Second Bruce A Reactor Cleared for Start-up*, 5 December 2003

[13] Atomic Energy of Canada Ltd, *Excellence Technology Partnerships*, Marcom Catalogue #1-2, D1498, April 2000

[14] Lamarsh J R and Baratta A J, op cit. p 163

[15] Nucleoelectrica Argentina (as of June 2004: www.na-sa.com.ar)

[16] Atomic Energy of Canada Ltd, *ACR-700 AECL's Next-Generation CANDU*, Marcom Catalogue #ACR-2, D1663, June 2002

[17] Atomic Energy of Canada Ltd, *AECL: The British Connection*, AECL Business Development Office, briefing sheet

[18] Hewitt G F and Collier J G, op cit. pp 129–130

[19] Atomic Energy of Canada Ltd, *ACR: Advanced CANDU Reactor*, AECL Business Development Office, briefing sheet

[20] Fischer U 1999 *Nuclear Engineering and Design* **187** 15–23

[21] Teichel H 1996 *Nuclear Engineering and Design* **165** 271–276

[22] Fischer M 2004 *Nuclear Engineering and Design* **230** 169–180

[23] Electricity Association, *Industry Facts* (as of June 2004: http://www.electricity.org.uk/default.asp?action = article&ID = 10)

[24] Davis M B, *France's and Europe's Next Pressurized Water Reactor*, Earth Island Institute (as of June 2004: http://www.earthisland.org/yggdrasil/reactor.html)

[25] Estève B, Presentation at International Congress on Advances in Nuclear Power Plants, Cordoba, Spain, 4–7 May 2003

[26] Uranium Information Centre, Briefing Paper 76, Melbourne, Australia (as of June 2004: http://www.uic.com.au/nip76.htm)

[27] Areva, Press Briefing, *Olkiluoto 3—un projet EPR clé en main en Finlande*, Paris, December 2003

[28] Framatome ANP, EPR Press Kit, January 2004 (http://www.framatome-anpcom)

[29] Nuclear Energy Institute, 'Up Front' briefing, *Consortia Apply to DOE to Test Licensing Process for New Nuclear Power Plants* (as of June 2004: http://www.nei.org/doc.asp?docid = 1184)

[30] Ministère de l'Economie des Finances et de l'Industrie, France, 2003, *Coûts de Référence de la production électrique*, DGEMP

[31] Teichel H and Pouget-Abadie X 1999 *Nuclear Engineering and Design* **187** 9–13

[32] Framatome ANP, Areva, Paris (as of June 2004: http://www.framatome-anpcom/)

[33] Greenpeace Sweden (as of June 2004: http://www.greenpeace.se/np/s/epr/)

[34] Global Chance, *Le réactuer nucléaire EPR: un projet inutile et dangereux*, Cahier No. 18, Meudon, France, January 2004

[35] Lamarsh J R and Baratta A J, op cit. p 143

[36] Argonne National Laboratory, Engineering Research, *Boiling Water Reactors* (as of June 2004: www.era.anl.gov/history/bwr.html)

[37] General Electric Power Systems, Press Release, *New President Named for GE Nuclear Energy*, San Jose, California, 10 March 2003

[38] Hewitt G F and Collier J G, op cit. p 46

[39] Hewitt G F and Collier J G, ibid. p 241

[40] Toshiba Corporation, *Construction Planning and Methods of the ABWR Plant*, brochure, Tokyo, Japan

[41] Department of Nuclear Engineering, University of California, Berkeley, research in advanced reactor designs (as of June 2004: http://www.nuc.berkeley.edu/designs/abwr/abwr.html)

[42] Nuclear Information and Resource Service, Washington DC (as of June 2004: www.nirs.org/globalization/GEGlobalization.htm)

[43] Framatome ANP 2002 *SWR1000 The Boiling Water Reactor with a New Safety Concept*, Offenbach am Main, Germany

[44] M Freedman, Testimony to the US House of Representatives, House Science Committee, Energy and Environment Subcommittee, Washington DC, 1 May 1996

[45] Uranium Information Centre, Briefing Paper 60, Melbourne, Australia (as of June 2004: http://www.uic.com.au/nip76.htm)

[46] Hewitt G F and Collier J G, op cit

[47] Simon R A and Capp P D, Proceedings of HTR-2002: Conference on High Temperature Reactors, Petten, The Netherlands, 22–24 April 2002, sponsored by the International Atomic Energy Agency

[48] Methnani M and Stanculescu A, *Status of High Temperature Gas-Cooled Reactor Technology*, paper presented at: the Workshop on Nuclear Reaction Data and Nuclear Reactors: Physics, Design and Safety, 25 February–28 March 2002, The Abdus Salam International Centre for Theoretical Physics, Trieste, Italy

[49] Past and present employees of the Fort St Vrain Generating Station, Website: *Fort St. Vrain Electric Generating Station* (as of June 2004: http://fsv.homestead.com/FSVHistory~ns4.html)

[50] Wang C et al, *Design of a Power Conversion System for an Indirect Cycle, Helium Cooled Pebble Bed Reactor System*, Proceedings of HTR-2002: Conference on High Temperature Reactors, Petten, The Netherlands, 22–24 April 2002, sponsored by the International Atomic Energy Agency

[51] Talbot D 2002 'The next nuclear plant' *Technology Review* **105**(1) 54–59

[52] Hoffman J M 2001 *Machine Design*, 27 September, pp 93–98

[53] Terry W K (editor), *Modular Pebble Bed Reactor Project*, Laboratory-Directed Research and Development Program FY 2001 Annual Report, Idaho National Engineering and Environmental Laboratory, December 2001 (Publisher's reference: INEEL/EXT–01–01623). NB Other annual reports in this series are also of relevance

[54] Zhong D et al, *China's HTR Test Module Project*, Proceedings of the Technical Committee Meeting on Response of Fuel, Fuel Elements and Gas Cooled Reactor Cores under Accidental Air or Water Ingress Conditions, Beijing, China, 25–27 October 1993, sponsored by the International Atomic Energy Agency

[55] Yuanhui X, *Chinese Point and Status—The HTR-10 Project and its Further Development*, Proceedings of HTR-2002: Conference on High Temperature Reactors, Petten, The Netherlands, 22–24 April 2002, sponsored by the International Atomic Energy Agency

[56] Yuanhui X, *High Temperature Gas-Cooled Reactor Programme in China*, Proceedings of the 4th Nuclear Energy Symposium—part of the series Energy Future in the Asia Pacific Region, Taipai, Taiwan, 15–16 March 1999

[57] Lyman E S 2001 'The pebble-bed modular reactor (PBMR): safety and non-proliferation issues?' *Physics and Society, American Physical Society*, **30**(4) 16–19

[58] Simon R A and Capp P D, *Operating Experience with the Dragon High Temperature Reactor Experiment*, Proceedings of HTR-2002: Conference on High Temperature Reactors, Petten, The Netherlands, 22–24 April 2002, sponsored by the International Atomic Energy Agency

[59] Japan Atomic Energy Research Institute, *Design of High Temperature Engineering Test Reactor (HTTR)*, September 1994, Report 1332 (as of June 2004: http://www2.tokai.jaeri.go.jp/httr/eng/jaeri_1332.html)

[60] Japan Atomic Energy Research Institute, HTTR website (as of June 2004: http://www2.tokai.jaeri.go.jp/httr/eng/index_top_eng.html)

[61] General Atomics, GT-MHR website (as of June 2004: http://www.ga.com/gtmhr)

[62] Gruppelaar H, Kloosterman J L and Konings R J M 1998 *Advanced Technologies for the Reduction of Waste* (Petten: Netherlands Energy Research Foundation)

[63] BNFL 1989 *Nuclear Fuel Reprocessing Technology* (London)

[64] Department of Energy, *A Roadmap for Developing Accelerator Transmutation of Waste (ATW) Technology*, Washington DC, October 1999 (DOE/RW-0519)

[65] Carleson T E, Chipman N A and Wai C M (eds) 1996 *Separation Techniques in Nuclear Waste Management* (Boca Raton, FL: CRC Press)

[66] Rao L and Choppin G R, 'North America' in *Separation Techniques in Nuclear Waste Management*, Carleson T E *et al* (eds) ibid. pp 239–249

[67] Bronson M C and McPheeters C C 'Pyrochemical treatment of metals and oxides' in *Separation Techniques in Nuclear Waste Management*, Carleson T E *et al* (eds) ibid. pp 155–167

[68] Worl L A 'Physical treatment techniques' in *Separation Techniques in Nuclear Waste Management*, Carleson T E *et al* (eds) ibid. pp 191–208

[69] M F Buehler and J E Surma 'Electrochemical processes' in *Separation Techniques in Nuclear Waste Management*, Carleson T E *et al* (eds) ibid. pp 91–107

[70] Santini P , Leímanski R and Erdos P 1999 'Magnetism of actinide compounds' *Advances in Physics* **48**(5) 537–653

[71] Crowley K D 1997 'Nuclear waste disposal: the technical challenges' *Physics Today* June 32–39

[72] Argonne National Laboratory, Environmental Assessment Division, *Human Health Fact Sheet: Technetium* (as of July 2004: http://www.ead.anl.gov/pub/doc/technetium.pdf)

[73] Koch L, Glatz J-P, Konings R J M and Magill J *Partitioning & Transmutation Studies at ITU*, ITU Annual Report 1999 (EUR 19054) p 4

[74] Gruppelaar H, Kloosterman J L and Konings R J M, op cit. p 13

[75] The Royal Society 1998 *Management of Separated Plutonium* (London)

[76] Uranium Information Centre, May 2002, Briefing Paper 35 (as of October 2004: http://www.uic.com.au/nip35.htm)

[77] *Management of Nuclear Waste*, Select Committee on Science and Technology, House of Lords, HL Paper 41, London (1999)

[78] Konings R J M and Haas D 2002 'Fuels and targets for transmutation' *C. R. Physique* 3 1013–1022

[79] Gruppelaar H, Kloosterman J L and Konings R J M, op. cit. p 35

[80] Minatom 2002 *Minatom of Russia 2002 10th Anniversary of Minatom of Russia*, Central Research Institute of Management, Economics and Information, Moscow, Russia

[81] Adamov E O, Ganev I, Lopatkin A V, Orlov V V and Smirnov V S 2000 *Nuclear Engineering and Design* **198** 199–209

[82] Adamov E, Orlov V, Filin A, Leonov V, Sila-Novitski A, Smirnov V and Tsikunov V 1997 *Nuclear Engineering and Design* **173** 143–150

[83] Minatom, Press Release, July 2000, *International Seminar—Cost Competitive, Proliferation Resistant, Inherently and Ecologically Safe Fast Reactor and Fuel Cycle for the (sic) Large-Scale Power*, Moscow

[84] Diggis C, *Minatom Jumps the Gun on Reactor Cooperation Program with US DOE*, Bellona, 30 May 2002 (as of July 2004: http://www.bellona.no/en/international/russia/npps/co-operation/24370.html)

[85] Bowman C D, Arthur E D, Lisowski P W, Lawrence G P, Jensen R J, Anderson J L, Blind B, Cappiello M, Davidson J W, England T R, Engel L N, Haight R C, Hughes III H G, Ireland J R, Krakowski R A, LaBauve R J, Letellier B C, Perry R T, Russell G J, Staudhammer K P, Versamis G and Wilson W B 1992 'Nuclear energy generation and waste transmutation using an accelerator-driven intense thermal neutron source' *Nuclear Instrumentation and Methods* **320** 336–367

[86] Klapisch R 2000 'Accelerator driven systems: an application of proton accelerators to nuclear power industry' *Europhysics News* **31**(6)

[87] CEA, *From the Critical Reactor to the Subcritical Hybrid System: Transmutation Tools*, Clefs CEA No. 46, Paris, France, Spring 2002

[88] Gudowski W 1999 'Accelerator-driven transmutation projects: the importance of nuclear physics research for waste transmutation' *Nuclear Physics A* **654** 436c–457c

[89] Nifenecker H, Meplan O and David S 2003 *Accelerator Driven Subcritical Reactors* (Bristol: IOP Publishing)

[90] Reactor Physics Group of the Laboratoire de Physique Subatomique et de Cosmologie, Grenoble, France (as of July 2004: http://lpsc.in2p3.fr/gpr/activites.html)

[91] French Law No. 91-1381 of 30 December 1991 (as of July 2004: http://www.andra.fr/interne.php3?id_article = 399&id_rubrique = 123)

[92] ENEA, European Technical Working Group on ADS, *A European Roadmap for Developing Accelerator Driven Systems (ADS) for Nuclear Waste Incineration*, Rome, Italy, (2001)

[93] Revol J-P 2001 'An accelerator-driven system for the destruction of nuclear waste' *Progress in Nuclear Energy* **38**(1–2) 153–166

[94] Rodriguez C, Baxter A, McEachern D, Fikani M and Venneri F 2003 'Deep-Burn: making nuclear waste transmutation practical' *Nuclear Engineering and Design* **222** 299–317

[95] Zerriffi H and Makhijani A 2000 *Nuclear Alchemy Gamble: An Assessment of Transmutation as a Nuclear Waste Management Strategy* (Takoma Park, MD: Institute for Energy and Environmental Research)

[96] US Department of Energy Nuclear Energy Research Advisory Committee and the Generation IV International Forum, *A Technology Roadmap for Generation IV Nuclear Energy Systems* (GIF-002-00), Washington DC, December 2002

[97] US Department of Energy, Office of Nuclear Energy, Science and Technology, *A Roadmap to Deploy New Nuclear power Plants in the United States by 2010, Volume I, Summary Report* (as of July 2004: http://nuclear.gov/nerac/ntdroadmapvolume1.pdf), Washington DC, 31 October 2001

[98] Lennox T, *Research & Development for Generation IV Systems*, Presentation at Universities Nuclear Technology Forum 2004, University of Liverpool, 19–21 April 2004

[99] Brundtland G (ed) 1987 *Our Common Future* (Oxford University Press, Oxford: The World Commission on Environment and Development)

[100] Thompson D (ed) 1996 *Pocket Oxford Dictionary*, 8th edition (Oxford: Oxford University Press)

[101] Uranium Information Centre, *Generation IV Nuclear Reactors* Briefing Paper 77, Melbourne, Australia, August 2003

[102] Brown P 'Sellafield shutdown ends the nuclear dream' *The Guardian*, Tuesday 26 August 2003

[103] Massachusetts Institute of Technology 2003 *The Future of Nuclear Power: an Interdisciplinary MIT Study* (Cambridge, MA)

[104] Website of Idaho National Engineering and Environmental Laboratory, Idaho Falls, Idaho USA (as of July 2004: http://energy.inel.gov/gen-iv/)

[105] Lockheed Martin, Ridgelines, *Ending the MSRE, 12 November 1998* (as of July 2004: http://www.ornl.gov/info/ridgelines/nov12/msre.htm)

[106] Patterson W 1999 *Transforming Electricity* (London: Earthscan)

PART III

NUCLEAR FUSION TECHNOLOGIES

Chapter 9

Fusion: an introduction

If truth be told, the separate nuclear communities advancing the civilian technologies of nuclear fission power and magnetically confined fusion power have for 50 years had something of a difficult relationship with one another. On the side of mutual affection is a combined early history with common heroes. Also both technologies have their roots in fundamental 20th century physics and to some extent a common heritage of big science. On the negative side the fusion community feel aggrieved when the public and policy-makers draw parallels between fusion and fission power generation. As the fusion community are keen to point out, their technology has no stored nuclear energy and no long-lived radioactive wastes. They regard it as unfair when they suffer by association with the key negative attribute of nuclear fission: public anxiety concerning safety and radioactive wastes. The European fission research community notes with some jealousy the extremely large sums that the nuclear fusion research community receives from Euratom (€750 million in the four year Sixth Framework Programme of the European Union) [1]. While in recent decades nuclear fission has largely failed to win the political argument in the west, the fusion community has maintained a remarkable talent for gaining the support of senior politicians. Some of these differences in terms of public and political support no doubt arise as a consequence of the differing position of the two technologies in their respective life cycles [2]. Nuclear fission is widely regarded as a mature technology while nuclear fusion electricity production is yet to be demonstrated. Fusion is still very much a research-led activity.

Nuclear fission may be summarized by noting that, along with supersonic passenger transport and space exploration, it represents one of the key signs of the power of engineers in the third quarter of the 20th century. The years after the allied victory in that most technologically determined of conflicts, World War II, can be remembered as the high watermark of uncompromising engineering-led design. The engineers were free to design and build technologies at the limits of the possible, without regard to cost, real utility, external impacts or the merits of alternative opportunities. In the past 25 years, however, engineers have been forced to adopt a more pragmatic outlook and to recognize that engineering excellence can be only one factor in the development of large-scale technology projects.

Perhaps the clearest manifestation of the new pragmatism in engineering is the creation by the Massachusetts Institute of Technology of an Engineering Systems Division. This recognizes the systemic aspects of engineering and the need for effective linkages with cognate disciplines such as law and economics. MIT has created the ESD to complement its world-leading engineering science departments. As previous chapters have argued, if nuclear fission is to have a renaissance then it will not be on the basis of 1960s style engineering-led design, but rather in the form of a pragmatic, engineering systems approach, balancing several key factors including total real cost, environmental impact and public preferences.

By contrast, through the second half of the 20th century nuclear fusion has remained the domain of the physicists. Its developers have regarded it as a series of experiments. It is a research challenge centring on the most fundamental properties of nature. However, just as nuclear fission needed to make transitions in its culture, so now must nuclear fusion. Fusion must now make the transition from a science to a true technology.

It could be regarded as odd to be discussing nuclear fusion in a book entitled *Nuclear Renaissance*. After all, things can hardly undergo a renaissance if they have never achieved viability. The lack of demonstrated viability should not be interpreted too harshly, however, for whatever the semantic details, nuclear fusion seems likely to play a significant role in the long-term future of nuclear power.

In these chapters we shall consider various approaches to the eventual generation of electricity using nuclear fusion energy. We shall emphasize the two communities of magnetic and inertial confinement fusion with their differing approaches, cultures and mutual aloofness. We shall not consider possible technologies lying outside these two paradigms. For instance shall not emphasize the interesting physics of muon-catalysed fusion as it is unlikely to produce electricity in foreseeable timescales. Such research is actively under way into the physics of the fusion of small muonium (a light muon substitutes for the proton in the nucleus) isotopes of hydrogen at the TRIUMF laboratory in Canada, at the ISIS facility of the Rutherford Appleton Laboratory in Oxfordshire, and in France and Japan as well. A large research literature is available for those with interests in this direction.

We shall not consider the controversial, widely debunked, and yet not wholly forgotten possibility known as 'cold fusion'. Even in the apparently unlikely event that there is anything of interest in such electrolytic experiments, it seems most unlikely that any significant contribution to global energy supply could result. Nor shall we consider the more tantalizing but similarly controversial reports of linkages between the phenomenon of sonoluminescence and nuclear fusion (so-called 'bubble fusion') [3].

In this chapter and the following chapter we shall emphasize the leading technology for fusion electricity production—the magnetic confinement approach. In chapter 11 we shall consider an independent approach based upon high-pressure and high-temperature compression of fusion fuels. This alternative approach is known as inertial confinement fusion.

In magnetic confinement fusion the fuel is introduced in the form of a gas of hydrogen isotopes. At the high temperatures involved the electrons and the nuclei dissociate. The electrons form a gas of free charged particles separate from, but intermingled with, a gas of the bare hydrogen nuclei. Two co-existing intermingled gases of charged particles are a special state of matter. It is this state of two co-existing charged gases that is termed 'plasma'. While on first impression one might take the view that this state of matter is unfamiliar, in fact we all have significant day-to-day experience of plasmas, and we probably tend to think of them as simply glowing gases. Flames, the workings of fluorescent light bulbs and neon streetlights, lightning, the sun, and other stars are all fundamentally forms of plasma. If plasmas are the fourth state of matter, then for many years they were the one that individuals could not easily manipulate. They were the most magical of phases, which one was fortunate even to observe. This aspect of natural plasmas is well reflected by the mysteries of one particular plasma: the Aurora Borealis. However, in recent years with the widespread availability of novelty lamps known as plasma spheres, ordinary people can finally play with plasmas at home.

Despite the fact that magnetically confined plasmas for energy production are only now on the cusp of becoming a technology, the study of plasma physics is far from a recent innovation. Plasma science had its roots before World War II in the study of astrophysical phenomena. After the war, however, two significant spurs were given to plasma physics independent from our main concern of electricity production. The first was the interest of the defence sector in plasma physics underpinning the development of thermonuclear weapons (i.e. the hydrogen bomb). The second was the interest of the electronics industry in low-temperature plasmas initially for thin film deposition for integrated circuit manufacture. Plasma technology has now been applied in a vast range of industrial sectors. Perhaps the most visible in everyday life is the growing prevalence of high brilliance flat plasma television displays. For a summary of the full range of plasma applications see Timothy E Eastman's excellent plasma science website *Perspectives on Plasmas* in which he surveys the industrial application of plasma science [4]. The bulk of the innovations in industrial plasmas have been relatively recent, although much of the fundamental science dates back to the 1970s. The roots of fusion power research, however, were planted right back in the 1950s and lagged only a few years behind the early efforts to produce fusion weapons systems. It is not the purpose of this book to provide an historical review of fusion research and experimentation. Readers seeking such insights are encouraged to consult C M Braams and P E Stott's excellent technical history of magnetically confined fusion research entitled *Nuclear Fusion* [5]. Here we shall look only at fusion's past in so far as such historical insights can assist us in understanding the context of the possible future technological developments of a nuclear renaissance.

Fusion is a process that generates nuclear energy from physical principles related to, but quite distinct from, the process of nuclear fission considered in chapter 3. As the names suggest, nuclear fission obtains energy through the splitting of large heavy atoms, while nuclear fusion produces far greater levels of energy through the joining together of small light atoms. The key to understanding all nuclear power is an appreciation of the binding energy of the atomic nucleus. As discussed in chapter 3 the nucleus consists of positively charged particles known as protons and electrically neutral particles known as neutrons. These are all held together in a compact space by a special force known as the **strong nuclear force**,[1] of which in day-to-day life we have no experience. Despite our lack of experience of the actual realities of nuclear physics, analogy with our day-to-day experience can be helpful in gaining an understanding of the issues involved. To appreciate **nuclear binding energy** it is useful to use Niels Bohr's analogy of 1917 that the atomic nucleus may be regarded as a liquid droplet, such as a droplet of water. Water droplets are held together by a weak and fluctuating electrical attraction between water molecules; this force is known as hydrogen bonding and we can regard it as being analogous to the strong nuclear force. The other property required to produce water droplets is surface tension and it is this that is the analogy of the nuclear binding energy. As a surface effect, surface tension becomes proportionately weaker the larger the droplet. Very large water droplets struggle to hold together and are easily encouraged to split into two smaller water droplets in which there will be more surface area and hence more surface tension. This process of an overly large water droplet being prompted to split into two more stable droplets is analogous to the process of nuclear fission discussed in chapter 3. At the other end of the size scale one can consider the emergence of droplets. As one considers ever-smaller droplets, at some level one is forced to concede that one is considering a small collection of molecules and not something that should be described by the macroscopic term droplet. A collection of molecules has no surface in the macroscopic sense, and hence no surface tension. Individual molecules have no incentive in energy terms to combine and nucleate a droplet. Typically for real droplets (and bubbles) nucleation actually occurs on a piece of dirt e.g. a smoke particle or at a defect on a substrate. If two small collections of water molecules can be

---

[1] As with the treatment of nuclear fission in chapter 3, technical terms key to an understanding of nuclear fusion are shown in bold.

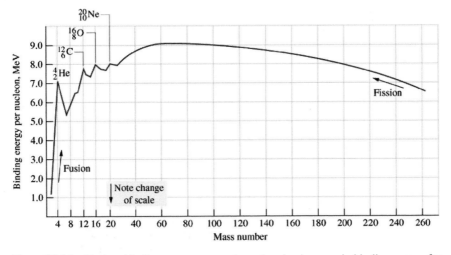

Figure III.9.1. Nuclear binding energy per nucleon showing increase in binding energy for processes of both nuclear fusion and nuclear fission (source: J W Hill and R H Petrucci, *General Chemistry* [6]).

brought together to fuse into a true droplet then the benefits of the creation of surface tension will occur. This is the analogy to the change in binding energy achieved when fusing two light nuclei. As figure III.9.1 illustrates, the change in binding energy between two nuclei of neighbouring sizes is much more dramatic as one constructs large nuclei from very small nuclei (fusion) than it is if one splits overly large nuclei (fission). This is the basis for the far larger amounts of energy released in a single nuclear fusion event than in an equivalent nuclear fission event. This is the fundamental reason why the hydrogen bomb is so much more powerful than the atomic bomb.

Famously nuclear power relies on Einstein's observation that mass and energy are equivalent ($E = mc^2$). This is as true for fusion as it is for fission. In both cases binding energy manifests itself in the form of measurable mass. That is, if one measures the masses of the constituents of the process before and after the nuclear reaction (adjusting where necessary for any relativistic effects owing to the high speeds of some of the components) then one finds that the components before the reaction have a higher mass than afterwards. This difference in mass is equivalent (via Einstein's equation) to the difference in binding energy before and after, and is equal to the energy released in the process.

In the early stages of any fusion power programme it seems likely that the nuclear reaction used would be that between two **isotopes** of hydrogen: **deuterium** (D) and **tritium** (T) (figure III.9.2). At temperatures of approximately 100 million degrees (in any units) and at pressures of approximately 8 atmospheres these two isotopes would be likely to fuse, forming helium and

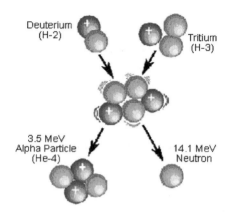

Deuterium
(H-2)

Tritium
(H-3)

3.5 MeV
Alpha Particle
(He-4)

14.1 MeV
Neutron

Figure III.9.2. The deuterium–tritium fusion process (source: Wikipedia [7]).

a neutron, and releasing energy sufficient that a fusion-based power station could be viable. Under these conditions none of the constituents is in atomic form. The hydrogen isotopes will have lost their electrons and likewise the helium nucleus produced will lack bound electrons.

In order for the deuterium and tritium ions to come close enough to fuse despite their mutual electrostatic repulsion, very high temperatures and high pressures are needed. The conditions above which sufficient fusion will occur for a net power output are defined by **Lawson's Criterion** which specifies that for D–T fusion the product of **confinement time** and **ion density** must exceed 10^{20} s/m^3 at a temperature of 30 keV (10 keV corresponds to roughly 100 million Kelvin; also see section III.9.1) [8].

A fast moving helium nucleus is conventionally known as an **alpha particle**. Such particles have been observed in the earliest studies of radioactivity. The D–T fusion process releases 17.6 MeV of energy. 14.1 MeV of this is given to the fast neutron produced. It is this fast neutron that is planned to provide the basis for electricity production. As neutral particles, the fast neutrons pass straight out of the plasma unaffected by the magnetic fields in the reactor and they impact the walls of the machine—the **blanket** (see section III.9.2). Here the neutron loses its energy through a series of collisions heating up the blanket assembly. It is this heat energy that could be used to make electricity. The remaining 3.5 MeV of energy passes to the helium nucleus (alpha particle). This charged particle remains in the plasma, depositing its energy in the plasma. If sufficient energy were passed to the helium nuclei then it is possible that enough heat would be retained within the plasma that no external heating at all would need to be supplied. The point at which this occurs is known as **ignition** and for many years it was assumed that a power station would necessarily need to operate in this state (see section III.9.4). These days, however, it is realized that ignition is not required and in many ways it would be positively undesirable.

Once the helium nuclei lose their energy they become a problem for the plasma. They inhibit future D–T fusion because they dilute the fuel and with time they cause a cooling of the plasma. Consistent with its status as the waste generated by the fusion process the alpha particle by-product has become known as helium ash. It is something that must be removed from the plasma, although it is important to stress that it is not a radioactive waste. The task of removing the helium ash is performed by the **divertor** at the bottom of the tokamak (see section III.9.2). The divertor also removes other charged impurities, such as carbon nuclei ablated from the walls of the toka-mak. Plasma research has shown that any sharp corners of graphite tiles sticking out from the smooth wall of the vessel will glow white hot and release carbon nuclei into the plasma—a process known as **carbon bloom**. As with helium ash these ions inhibit the process of nuclear fusion and must be removed (see section III.9.2).

While the simplistic description renaissance may strictly be inappropriate for fusion, it would equally be wrong to infer that the development of fusion has been one slow climb from the earliest scintilla of knowledge to a technol-ogy on the cusp of demonstration. In fact the history of fusion energy research has been something of a roller-coaster ride with enormous highs of optimism and deep lows of disappointment and frustration. Key elements of this experi-ence include Sir John Cockcroft's late 1950s announcement to a journalist that he was 90% certain that neutrons emitted by the Harwell Laboratory flagship Zero-Energy Thermonuclear Assembly (ZETA) were the expected product of a nuclear fusion reaction. The euphoria of this moment for Cockcroft and his team was short lived, as it was soon found in May 1958 that fusion processes were not producing the measured neutrons at all [8].

Through the 1960s fusion researchers had to grapple with the fact that fusion plasma confinement is far from straightforward. It became increasingly clear that plasma instabilities were serious, ubiquitous and very hard to minimize and control. It was never the case, however, that high-temperature plasmas were so tricky as to force the community to give up their attempts. Rather, the stable behaviour of hot plasmas in magnetic fields, while largely elusive, was felt at every point to be almost within the scientists' grasp. While the researchers remained optimistic, their paymasters were becoming less so. In 1968 the Lighthill Report recommended, as a consequence of the high level of technical difficulties being encountered, that the UK magnetically confined fusion power programme should be shut down.

The increasing anxiety of western fusion researchers and the despon-dency of the policy makers were largely lifted in 1968 with reports from two young British physicists, Derek Robinson and Nicol Peacock, working in the Soviet Union. Robinson and Peacock, together with colleagues Mike Forrest and Peter Willcock, had measured plasma temperatures inside the Russian T3 **tokamak** device using light scattering techniques. Their work confirmed impressive Russian claims for the tokamak that had previously not been believed in the west. Remarkably, these two researchers were working at the Kurchatov Institute in the Soviet Union in the dark Cold War years before détente. Robinson and Peacock's measurements confirmed that the Russians had indeed achieved a breakthrough in magnetically confined fusion with their tokamak concept (see section III.9.1). This good news gave fresh momentum to western fusion efforts and talk of giving up receded in the face of a new wave of optimism.

In addition to bringing the Russian innovations to the attention of western researchers Robinson and Peacock were also an early part of a trend that would last for a quarter of a century, namely that magnetic confinement fusion research could be a vehicle for technological cooperation between the two power-blocs of the Cold War. Through the 1970s and 1980s fusion would provide a welcome channel for technological dialogue across the Iron Curtain.

The 1970s were dominated by the construction of three large tokamak devices: one in the United States, the tokamak fusion test reactor (TFTR) at Princeton University, the JT-60 in Naka in Japan, and the largest of the three the Joint European Torus or JET built at Culham in England. This tendency to increasing size is important for tokamaks, as with increased size comes greater plasma stability (see section III.9.1).

Through the 1980s the first waves of experiments were performed on the new big tokamaks. These concentrated on the problems of heating the plasma. This challenge turned out to be far greater than had been anticipated. Ohmic heating induced in the plasma was found to be far from sufficient. In addition significant difficulties were experienced in trying to keep the plasma clean, leading to significant redesigns to the torus walls [9]. During the 1980s a series of key plasma physics experiments were performed on D–D reactions. These demonstrated that the key plasma properties could indeed be controlled, but in the absence of the use of radioactive tritium only very small amounts of actual fusion were occurring.

The early 1990s saw significant movement on both a policy front and a technical front. Both JET and TFTR started a series of D–T reaction experiments that further demonstrated the validity of the key assumptions regarding magnetically confined nuclear reactions. In 1997, for instance, JET produced fusion power of 16 MW using D–T reactions at the peak of a pulse lasting two seconds (see figure III.9.3). As such this equalled 64% of the heating power needed to heat the plasma, and JET was said to have an achieved 'Q-value' of 0.64. The state at which the total fusion energy created equals the heating energy supplied is known as 'breakeven' and denoted by $Q = 1$. It is important to note, however, that roughly 80% of the fusion energy is carried away by the neutrons as they immediately leave the plasma. This energy release makes a fusion power station possible. Given, however, that at breakeven 80% of the fusion energy leaves the plasma, it is inevitable that a reactor at breakeven would in fact cool and

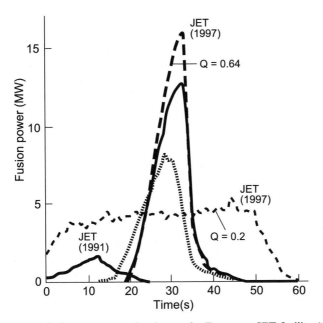

Figure III.9.3. D–T fusion energy production at the European JET facility (source: JET EFDA).

cease to produce fusion energy. A reactor running at $Q = 5$, or above, will leave sufficient heat in the plasma to maintain stable operations without the need for the external injection of heat.

Meanwhile a parallel policy development occurred in the mid 1980s when Soviet General Secretary Mikhael Gorbachev proposed to US President Ronald Reagan that the two superpowers collaborate for the benefit of all mankind on a next generation tokamak, the International Thermonuclear Experimental Reactor (ITER). A four-way collaboration involving the European Union and Japan was formally agreed in 1987. The late 1980s were heady days for fusion. The experiments were difficult, but going well, and fusion scientists had the support of the world's political leaders.

What fusion scientists could not have foreseen was the downside to follow from having linked their ambitions to geopolitics. Fusion science would be affected by changing international relations, and the changes were to be profound. The turn of the decade from the 1980s to the 1990s was a turbulent time. In November 1989 the Berlin Wall came down and in December 1991 the Soviet Union ceased to exist. The mutually beneficial relationship between fusion research and superpower politics was over. The Russian Federation replaced the Soviet Union as an ITER partner, and the Russian economic position worsened. Despite the difficulties, Russia continued to honour its commitments to the ITER vision and maintained a steady flow of high quality scientific and engineering contributions. Through the turbulence, the fusion community pressed ahead with its grand ambitions, almost oblivious to the economic and political realities of their situation. In 1998 the scientists presented their proposal known as the ITER Final Design Report. Their request was for between $5.2 billion and $5.7 billion. The collapse of the Russian economy and the failure of the US particle physicists attempt to get their $10 billion machine built in the early 1990s (the Superconducting Super Collider or SSC) should have warned the fusion community that obtaining the funds for ITER was not going to be easy.

Having started out as a bilateral initiative of the Soviet Union and the United States, by the end of 1998 ITER faced the ironic situation of the Russian contribution seeming to be in jeopardy and the US having formally withdrawn from the project. The US could not stomach the price tag associated with the version of ITER under discussion in the late 1990s. Furthermore some US fusion academics had cast serious doubt on the wisdom and cost effectiveness of that ITER proposal. Two theorists from the University of Texas at Austin, William Dorland and Michael Kotschenreuter, caused a storm in 1996 by calculating that turbulence would cause the tokamak to release its heat before ignition would occur [10]. As ignition was a fundamental aspect of the original ITER design, these criticisms were powerful. Gradually, however, the weight of scientific opinion swung into place behind the ITER concept and the power of Dorland and Kotschenreuter's doubts diminished.

Experimental results also intervened to keep ITER science alive. Since the early 1980s experiments had revealed the existence of **transport barriers** at the outer edges of the plasma. These act as sources of plasma stability and self-generating heat insulation for the plasma. Prominent among the transport barriers is the now much-studied **H-mode**. A big question for ITER was how much heating would be required before the stabilizing H-mode would appear. As theoretical modelling improved, the H-mode seemed to be an ever more easily achievable goal.

In the mid 1990s a tussle developed between Congress and the US Department of Energy over the costs of the ITER project. The sticker-shock was not helped by MIT fusion researchers who estimated that just the ITER design costs could be as high as US $1.2 billion [11].

Following the election of George W Bush as President of the United States in 2000 and the tragic events of 11 September in the following year, geopolitics has shifted once again, and in many ways in favour of fusion research. Energy security has grown to complement climate change as key political issues of the moment. The credentials of fusion in these areas are impeccable and on 30 January 2003, US Energy Secretary Spencer Abraham announced that the United States was to rejoin a reduced-scale ITER collaboration. The same month China entered the consortium, followed in June by South Korea. Fusion was back on track.

Despite the waxing and waning fortunes of fusion research over the past few decades, its fundamental attributes have remained unchanged. Magnetic fusion power would be a clean form of large-scale base-load electricity generation. The D–T fuel cycle yields no radioactive wastes and if appropriate materials are used in the reactor construction then there is no long-lasting radioactive decommissioning legacy. The operation of a fusion power plant would emit no greenhouse gases such as carbon dioxide or any other pollutants. If the future of electricity generation and transmission continues to rely on large-scale central generation and national and international transmission infrastructures, then it would seem that nuclear fusion power will be ideally placed to match the demands of the future.

The inherent safety of fusion is also a fundamental positive attribute of the technology. A fusion reactor operates in many ways like a gas boiler. Fuel is basically continuously supplied in small amounts in order to maintain the energy production process. At any one instant a fusion power station would only contain enough fuel for a few seconds, or at most minutes, of operation. Despite the large volume of a reactor tokamak (fractionally larger than ITER's 1000 cubic metres) the amount of fuel contained would be insufficient to cause harm off-site even in the event of a complete and very rapid fusion reaction. The chemical energy available from the plasma returning to hydrogen atoms and combusting with accidentally introduced oxygen would be utterly negligible as the density of the plasmas is very low.

The lack of substantial stored energy in a fusion system is in contrast to the properties of two mature, large-scale, non-polluting electricity generating technologies—hydroelectricity and nuclear fission. Conventional large-scale dispatchable hydroelectric systems involve dams holding back substantial volumes of water. Current nuclear fission systems all rely on a critical mass of fissile material in order to maintain the chain reaction necessary for power production. Both hydroelectricity and nuclear fission power plants have to plan extensively against the risk of catastrophic failure. Historically this has been planning to avoid an accidental and uncontrolled release of some of their stored energy. The high levels of stored energy are fundamental to the operation of these two technologies competing with fusion. All three technologies have the common attributes of low emissions, base-load efficiency and high capital costs. Recently attention has started to shift to the vulnerability of fission power plants and dams to terrorist attack. If this policy attribute continues to be of importance fusion's relative merits can only improve.

While fusion benefits from the lack of substantial stored energy in the system, it would be too strong to say that there are no health and safety issues to be considered. For instance, a rupturing and venting of the pressure vessel could release radioactive tritium gas and radioactive graphite dust arising from the degradation of the blanket during operations. Such releases would be of serious concern for worker safety, but could easily be contained in an outer containment building similar to that incorporated routinely into nuclear fission plants. Such containments would also be able to cope well with the possibility of a fire in the graphite blanket of the reactor.

A fusion power plant in operation would have similar safety issues to a conventional combined cycle gas turbine power plant or a coal-fired plant. While there is a need for good design and professional operational procedures to ensure worker safety, the possibility of an accident occurring that could have off-site safety consequences would seem to be vanishingly remote.

In addition to its environmental and safety benefits, nuclear fusion also has a great advantage with regard to the security of the electricity supply. The first wave of fusion power stations will use deuterium and tritium fuels. Deuterium is available in unlimited quantities from seawater. It is extracted as deuterated, or heavy, water. It is easily obtained by vacuum distillation from natural water and can therefore be purchased at modest cost [12]. As a short-lived (half-life 12.3 years) isotope, tritium is not available naturally. It is available from fission reactors or from nuclear accelerator laboratories as a by-product of operations. In a fusion power station tritium could be produced by the neutron bombardment of lithium. For information on US tritium production see Charles D Ferguson's article in the March/April 1999 issue of *The Bulletin of the Atomic Scientists* [13].

As the primary fuels for fusion power are available to all advanced economies domestically it may be regarded as an entirely domestic electricity generation technology free of the risks of geopolitical instability affecting its upstream fuel cycle.

Magnetic confined fusion, with its fuel cycle safe from international turmoil, and its research history dominated by examples of collaboration between Cold War adversaries, represents a technological challenge almost completely insulated from military concerns, the only area of direct overlap between magnetic confinement fusion research and nuclear weapons development being a familiarity with the handling of large quantities of radioactive tritium. The technologies and discoveries associated with magnetically confined fusion are conventionally regarded as being far from any technologies of direct interest for weapons and defence. This perception leads magnetically confined fusion research to be especially attractive to those researchers who wish to have nothing to do with military matters.

While substantial fusion research activities exist in countries such as Japan and Germany with no nuclear weapons programmes, it is also the case that several nuclear weapons states (such as the United States, Russia, France and the UK) have active magnetically confined fusion research communities. In these countries public funds support two parallel and almost entirely separate fusion research communities. One is dedicated to magnetically confined fusion research for electricity production and the other is dedicated to inertial confinement fusion research for thermonuclear weapons. While there are some research groups in academia with links to both communities, generally there is a clear separation of the civil and military fusion programmes. The personnel involved differ, the politics differ and the cultures of the two research communities differ. Both communities exist separately and have tended, until relatively recently, to remain aloof from one another.

The more weapons-related technologies of inertial confinement fusion and the prospects for the generation of electricity from such technologies will be considered in some detail in chapter 11.

Recognizing the possible lost opportunities in having two almost completely separate large-scale fusion research communities on the government payroll, policy makers in the US have in recent years encouraged a greater level of dialogue between the magnetic and the inertial confinement communities. In 1990 the Fusion Policy Advisory Committee in the US recommended that inertial and magnetic fusion research for energy production be pursued in parallel, but substantial communication between the two fusion communities did not get under way until 1999. The breakthrough was a two-week meeting in Snowmass, Colorado, in July 1999 organized by Hermann Grunder and Michael Mauel. Participants were surprised by the level of common concerns encountered and by the extent to which they faced common challenges [14]. For instance, the magnetic confinement community were surprised to hear helpful suggestions from the inertial confinement community concerning the problems of materials stability in plasma-facing components (see section III.9.2). The inertial confinement researchers revived the idea that, rather than using solid materials such as graphite, the magnetic confinement community might consider a flowing liquid as the plasma-facing material—liquid lithium was especially recommended. This would cope much better with the bombardment by highly energetic particles, as the movement of the atoms of the liquid would immediately heal structural damage. Furthermore such a blanket layer would have the benefit of producing tritium for the fusion reaction as a result of nuclear reactions between the fusion neutron flux and the lithium. Recent small-scale experiments regarding the use of liquid plasma-facing elements in magnetic confinement fusion devices have yielded positive results [15].

In the US the context for the 1999 Snowmass discussions had been the turmoil in the magnetic confinement fusion community following the US 1998 decision to withdraw funding from the original ITER collaboration. The need for the magnetic confinement community to think radically and to innovate was intense. In that environment the small possibilities of synergies with inertial confinement research were worth the trip to Colorado. Meanwhile in the UK, however, the magnetic confinement community, while worried for their future, had not been subjected to the same level of pressure as their US colleagues, and UK attempts to build synergies between the military and civilian programmes have been less intense. Aversion on the part of magnetic confinement researchers to dealing with their military peers is more than simply a reflection of personal politics; it is also a self-interested position. Any potential benefits of insights gained must be balanced against the drawbacks of having the peaceful credentials of magnetic confinement fusion eroded in public perception. The UK fusion review of 2001 is sensitive to political concerns regarding defence synergies. The review reports that the UK is right to exclude a defence link from its rationale for fusion energy research. However, one small exception is that it recommends that the civil fusion research community consult with nuclear weapons researchers in the area of tritium removal. Such expertise would be useful for the safe decommissioning of the JET facility and it would seem foolish for the civil community to rediscover existing knowledge in an area where the civil programme has nothing to contribute to military developments [16]. For further discussion of military–civilian synergies in fusion see chapter 11.

One of the great strengths of the magnetic confinement fusion community has been their sensitivity and responsiveness to national and international politics. Whether the issue is détente, energy security or the environment, fusion has rightly been able to sell itself to policy makers as part of the solution. The linkages of fusion with the issues of the day go so far as to include the current politics of globalization. Fusion research benefits enormously from the economies of scale that can be achieved when countries collaborate. The ITER project is a modest cost for each of the participating countries and yet the total amounts of resource devoted to the task are substantial. Fusion research is inevitably mutually reinforcing with the trend to socio-economic globalization. In contrast, many renewable energy technologies are small-scale and well suited to local management. Many single-issue political questions are not in fact isolated issues but rather are mutually inter-linked. The issues of nuclear fusion research allow one to see particularly clearly the linkages between the politics of globalization and the politics of energy. The linkages between fusion and globalization are examined sociologically in a paper by Bechmann and co-workers [17].

Consideration of fusion and globalization may for some readers conjure up a whole host of negative associations. Fusion technology might be regarded as a technological step towards global homogeneity and away from local self-determination. Fusion may therefore be regarded as being imposed rather than chosen. In fact much of the recent evidence points in another direction—towards international generosity.

China has a growing interest in fusion technology to support its long-term electricity needs without continued reliance on environmentally harmful coal mined in geographically awkward places. China has chosen to expand its research base in fusion and has been able to do so via the transfer of second-hand research machines from other countries. For instance, China's HL-2A machine is the old German Asdex tokamak transferred to China and reconditioned. Similarly China's HT-7 machine is a rebuild of the Russian T7 tokamak. Rather than fusion power being imposed on a reluctant China, that country has chosen to improve its capacity and high-level skills in this important technological area and it has been assisted in this policy by the generosity of the international fusion community. China's future plans are also internationally collaborative. It is currently constructing a brand new machine HT-7U in collaboration with international partners. On 23 January 2003 *Nature* reported that China would be joining the ITER collaboration. The magazine reported that China's Minister of Science, Xu Gusnhua, had announced, 'China intends to make a major contribution to the project in the form of material or funding' [18]. China's involvement in fusion is serious and a key indicator of the positive contribution of fusion to international relations in high-technology areas.

Section III.9.3 discusses how the many decades timescale of fusion research for electricity production has caused policy difficulties in recent years. These concerns led policy makers to ask what other technological benefits might already have emerged from the fusion research programme by way of spin off. Stephen O Dean provided one of the first assessments of such spin-off application in 1995 [19]. He catalogued a wide range of technologies and materials whose developments have been assisted by research originally undertaken for fusion energy purposes. His list is similar in scope to that of the *Perspectives on Plasmas* website considered earlier in this chapter [4]. By the end of the 1990s the interest of policy makers in fusion spin-offs had grown considerably. The somewhat vague linkages between plasma research and research in other areas was no longer considered sufficient to demonstrate the benefits of fusion. What were needed were examples of direct spin-off where a new product could be linked unambiguously to developments in fusion research.

Sensitive to these issues UKAEA Fusion in Britain established a Fusion and Industry initiative with a website (as of July 2004: http://www.fusion.org.uk/industry/) and a regular newsletter entitled *Fusion Business*. This newsletter highlights examples of specific products ranging from novel city buses to lamps for vehicle, traffic and rail signalling. Each technology has direct links to fusion research. In November 2001 the British Department of Trade and Industry (DTI) published a major two-part review of the UK fusion programme. The management consultancy firm Arthur D Little undertook the first part of the review [16], and considered the management of the programme. A second separate part of the review concentrated on the scientific quality of, and prospects for, British fusion research.

The management review stressed the industrial linkages of fusion research. These include the research overlap areas considered previously, together with quantitative finance. Quantitative finance is the use of advanced mathematical and physical modelling techniques to financial markets. These techniques are particularly useful in the area of risk management and hedging. For instance, a particularly rich synergy between plasma physics and quantitative finance occurs in a common interest in the solution of the Fokker–Planck equation [20].

The DTI management review of fusion considered direct legal contracts with industry in support of fusion research and found that for large companies these links were of minimal importance. However, for smaller companies they could be extremely important. The review gave some significant consideration to technological spin-off into other industrial sectors. It noted that for most of the history of British fusion research technology spin-off had been given no emphasis and that opportunities had therefore been missed. The review strongly endorsed the need to manage spin-offs better and noted the recent moves by the UKAEA in that direction.

While in recent years the European fusion community has moved to build stronger links with industry, there is a second strategic decision that remains in place. For decades the British and other western European fusion communities have resisted any attempts to distract them from the focused goal of eventual commercial electricity generation. This is despite the fact that magnetic confined fusion technology has the potential, if so directed, to yield other socially beneficial outputs on shorter timescales. These intermediate goals would make direct use of either the 14 MeV neutron flux from a D–T fusion reaction or the heat energy arising in a blanket.

The aversion of the fusion energy community to intermediate technological applications, however, goes beyond a simple concern for the focus and momentum of the programme. It includes a desire to maintain a clear separation between nuclear fusion energy research and the problematic issues of nuclear fission. These links are particularly strong when one examines possible applications of high fluxes of 14 MeV neutrons. The Russian fusion community did not share the opinion of their western colleagues that such options represented more of a threat than an opportunity and they have devoted significant effort to fission/fusion hybrids. The Soviet Union even hosted a joint USSR–USA workshop on the topic in 1977 [21].

In a landmark 1999 article in *Fusion Technology* Wallace Manheimer of the US Naval Research Laboratory advocated a return to idea of a fission/fusion hybrid [22]. He reminded the fusion community of the political need to demonstrate socially useful results on timescales shorter than those of the conventional strategy. He points out that a hybrid fission/fusion reactor could be constructed around a smaller and far cheaper tokamak than the ITER machine. An effective hybrid could be based upon a tokamak similar in size to existing large-scale devices, such as JET and JT-60. Manheimer's suggestion is that the primary source of energy in the system would arise from the interaction of fusion 14 MeV fast neutrons with either thorium-232 or uranium-238 in a breeder blanket. With such blankets the machine would be able to breed fissile uranium-233 or plutonium-239 respectively. In this way the fusion hybrid could act not only as a power station in its own right, but also as a source of fuel for a fleet of conventional light water fission reactors. For instance, a hybrid with a thorium blanket could provide fuel for five light water reactors [22]. Manheimer does not go on to stress an important added benefit of this approach, which would be to partly separate fusion energy supply from the increasingly intolerant electricity generation industry. As with nuclear hydrogen discussed in section I.3.4, the fission fusion hybrid breeder reactor could be a technology more tolerant in terms of reliability and maintenance requirements than a base-load power station. A fission/fusion hybrid breeder would raise similar proliferation issues to a conventional fast breeder reactor (see chapter 7). However, as Manheimer points out, any fusion facility will require international nuclear safeguards to prevent the surreptitious production of fissile materials in the reactor blanket [22]. Manheimer extends his advocacy of fusion hybrids in a second article published in 2000 [23]. Advocates, such as Manheimer, believe that viable fusion hybrid power plants could be operational and cost effective far more quickly than conventional pure fusion tokamaks. As such, this intermediate utility would demonstrate the importance of nuclear fusion within a nuclear renaissance and could act as a pull on further fusion development.

Fission/fusion hybrid technology would also raise the possibility of an even shorter-term benefit in the area of fission nuclear waste management. Such a technology has the potential for application in the transmutation of nuclear wastes and surplus radioactive materials (cf. chapter 7). In the competition for public resources for new long-term energy technologies there are inevitable elements of competition between the magnetic confinement fusion community and those advocating accelerator-driven systems (ADS) for sub-critical fission power reactors. Synergies between ADS physics and fusion engineering are discussed in section III.9.3 in the context of materials irradiation for the fusion fast track. The idea that fusion and ADS also have similar possibilities in the area of nuclear waste transmutation further adds to the potential synergies. It is interesting, therefore, to note that the two research communities (ADS and fusion) have adopted very different attitudes to intermediate goals in their programmes. It is further of interest to note that the links between the two research communities are not as great as they ideally should be. In large part this seems to arise from the long-standing desire of the European fusion community to stand apart from the problematic world of nuclear fission. While this strategy has, so far, served the fusion research community well, it would seem that the time is fast approaching when the fusion community should re-examine this decision if it is to be an early contributor to the nuclear renaissance. As the prospect of real electricity production becomes closer

Plasma performance: beta and the triple product

The fusion physics community are rightly proud of the progress made in fusion plasma physics. In fact so much progress has been made in the fundamental plasma physics that the technical challenges facing the development of a functioning fusion power station are more to do with engineering and economics than they are to do with physics. Two parameters are used to benchmark progress in fusion research. One is the straightforward energy multiplication factor Q defined earlier. The other is the parameter β, defined as the ratio of the conventional pressure in the plasma p to the so-called 'magnetic pressure' or energy density, which scales as the square of the applied magnetic field B. In SI units β is defined by the equation

$$\beta = \frac{2\mu_0 p}{B^2} \qquad\qquad (\text{III}.9.1)$$

where μ_0 is the permeability of a vacuum.

For a stable plasma, first impressions would imply that the maximum value that β could hold would be $\beta = 1$, as the plasma pressure cannot exceed the applied magnetic pressure or the plasma would not be contained. In fact, the confining magnetic field can be larger than the value B appearing in equation (III.9.1) because in many cases that magnetic field is the nominal value applied or measured. It is not necessarily the actual value of the field at the key point of interest in the plasma. For such reasons, values for β greater than unity are sometimes reported [26].

The parameter β should not be confused with another related parameter the normalized β, β_N, or the Troyon factor. This is the value of β divided by the normalized plasma current I_N (i.e. unit-less and holding a maximum value of unity at design limits). As such, β_N is a pure measure of a given plasma (as limited by the machine design) and not of the operational efficiency of the machine. In practice β and β_N are not confused owing to their differing magnitudes. The ratio β might typically be expressed as, say, 7%, while the Troyon factor might have a value of 3.5 [27].

As a measure of performance, the higher the value of β the better the performance of a fusion machine.

Another important parameter is the triple product. This is the product of three terms the plasma density (conventionally n), the plasma temperature T and the energy confinement time τ_E. The triple product is closely related to the Lawson criterion described earlier. The Lawson criterion gives the condition under which sufficient fusion will occur for a net power output. As fusion rates depend on plasma temperature, the triple product allows us to define a region in which ignition can be expected occur: $nT\tau_E = 3 \times 10^{21} \, \text{keV m}^{-3} \, \text{s}$ [28].

The triple product achieved in different tokamaks is also a key measure of machine performance. The fusion community are rightly proud that the rate of improvement of this key measure of plasma physics exceeds the rate of improvement in integrated circuit computer chip transistor density and the improvement of particle physics experiments (figure III.9.4).

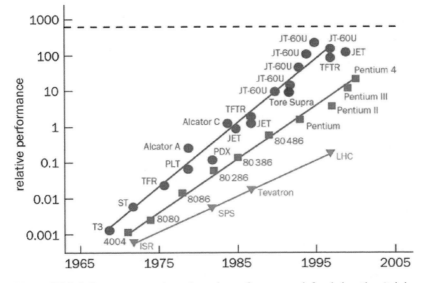

Figure III.9.4. Improvement in tokamak performance defined by the 'triple product' as compared with progress in computer electronics and particle physics. The dashed line denotes the performance expected of the ITER machine (source: Hoang and Jacquinot [28]).

there will be a need for greater pragmatism and for a reassessment of the community's culture and sense of self.

The range of non-electrical benefits of magnetic confined fusion is potentially very large and they have been summarized neatly by the ARIES team in the United States [24].

Such considerations force one to ask 'why, really, do we do these things?' One is led naturally to the conclusion that there is far more at stake than simply a desire to produce limitless environmentally benign electricity. Nations involve themselves in fusion research for a host of reasons including a desire to ensure globally competitive technical competencies. The magnetic confinement fusion community are rightly proud of their civilian peaceful credentials. It is nevertheless the case that countries participating in large-scale fusion research assist their ambitions to be one of the great powers with all of the civilian and military responsibilities that this entails. Furthermore one should not underestimate the importance of national prestige that emerges from participation in flagship technological projects. The path to a fusion power station, while clearly utilitarian in some respects, is in addition part of a long tradition of building grand designs which stretches from the pyramids, through the medieval cathedrals, to modern space stations. These aspects are an important driver of the process and should not be neglected when asking 'why do we do these things?'

Such sociological pressures can touch upon interesting differences in national attitudes and cultures. In his influential history *The Audit of War*, Corelli Barnett points to the centuries-long esteem in Britain for the lone gentleman innovator and disdain for the technical school with its emphasis on the efficiency of the organization [25]. Consideration of the needs and challenges of the big tokamak projects leads one to infer that fusion represents a good illustration of the problems Britain faces with such cultural legacies. However, the story of the spherical tokamak discussed in chapter 10 reminds us of the flipside of such cultural coins. The spherical tokamak represents exactly the kind of radical unconventional innovation that highly individualized societies such as the United Kingdom and the United States are known for.

Magnetic confined fusion is big science and big technology in every sense. The political issues raised by the technology and its research needs are complex and multi-faceted. This complexity is one of the key benefits of the technology as the lessons learned in international collaboration and project management alone will be of key benefit to the world in the decades to come.

III.9.1   Fundamentals of fusion: the tokamak

The word 'tokamak' is a contraction of the Russian phrase *toroidalnaya kamera ee magninaya katushka* meaning toroidal chamber and magnetic coil. The tokamak is the lead design concept for first generation fusion power production (see figure III.9.5). It relies on strong magnetic fields to confine the hot plasma in a toroidal geometry. The torus (or ring doughnut) shape for plasma confinement dates from the earliest days of magnetic confinement fusion research. In fact the 1950s ZETA machine at Harwell was a form of torus. What makes a tokamak special is the set of magnetic fields applied to the plasma held within the torus.

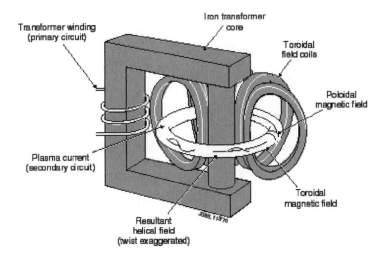

Figure III.9.5. Schematic representation of a tokamak showing its conceptual basis as a transformer with a multi-turn primary circuit and a single turn secondary circuit made up of the plasma itself. Fundamental to the tokamak concept is the use of large toroidal field coils providing plasma stability (source: EFDA JET).

Central to the concept of a tokamak is the idea that the device is a large transformer. A transformer is an electromagnetic device that can convert high-voltage alternating currents into low-voltage alternating currents of the same frequency but of higher current. Conventionally such transformers are symmetric and could in principle be reversed in order step up a voltage rather than to step it down. The trick behind a transformer is the number of windings of the magnetic coil through which the currents are required to pass on both sides of the transformer and the magnetic coupling of the two, for instance via a common iron core. A perfectly efficient transformer with 100 turns on the primary and only 10 turns on the secondary would step voltage down by a factor of ten and current up by the same factor. What is special about a tokamak is that the secondary winding of the transformer is not a multiple winding of copper wire, but rather is a single winding comprising the charged plasma itself. Any fluctuation in voltage and current applied to the primary coil is received via the iron core passing through the centre of the torus and transmitted in stepped up current to the plasma. As the plasma contains both negative electrons and positive nuclei two separate motions happen simultaneously. The electrons start to accelerate in one direction toroidally (i.e. around the torus in the horizontal plane) while the nuclei start to move in the opposite direction. The tokamak has some outer poloidal coils to position and shape the plasma (not shown in figure III.9.5), but the main poloidal effects arise from the toroidal flow of the current in the plasma, which in turn is driven by the action of the

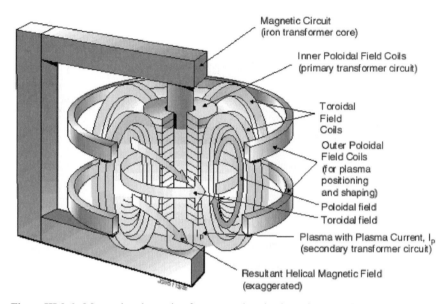

Magnetic Circuit
(iron transformer core)

Inner Poloidal Field Coils
(primary transformer circuit)

Toroidal
Field
Coils

Outer Poloidal
Field Coils
(for plasma
positioning
and shaping)

Poloidal field

Toroidal field

Plasma with Plasma Current, I_p
(secondary transformer circuit)

Resultant Helical Magnetic Field
(exaggerated)

I_p

Figure III.9.6. Magnetic schematic of a conventional tokamak. Note the inner and outer poloidal field coils. The net consequence of both the transformer action and the fields from the toroidal coils is for the plasma current to circulate around the torus in a horizontal plane. Individual ions follow helical paths as they orbit around the loop of the torus. In most real tokamaks the primary transformer coil winding (a solenoid) is actually located in the centre of the torus itself and not externally to the torus as shown in figure III.9.5 (source: EFDA JET).

transformer. Key to the stability of plasma flows within the tokamak is a special set of coils, known as toroidal field coils. It is the strong toroidal fields that were the key innovation of the tokamak when compared with earlier concepts for fusion. The strong toroidal fields provide forces on the flowing plasma that causes it to circulate helically around a path running around the centre of the torus. The positively charged nuclei flow helically in one direction while the negatively charged electrons flow helically in the opposite direction.

The plasma current in a tokamak gives rise to another advantage over competing geometries and that is Ohmic or resistive heating. This is exactly analogous to the resistive heating of a 1 kW bar in an old-fashioned electric fire. When the plasma is cold and viscous much of the energy put into the primary of the transformer is used in getting the plasma current going. The losses in terms of poor plasma circulation turn out to be a bonus as the energy not transferred to the plasma circulation has in fact gone to the warming of the plasma. As the device warms up, more and more of the electrical energy applied to the primary is successfully transferred to

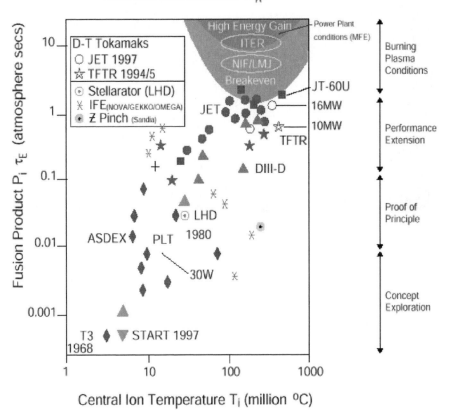

Figure III.9.7. Fusion triple product: an indicator of fusion performance in worldwide tokamaks (see box: *Plasma performance: beta and the triple product*). Note the logarithmic scale on the left, and the history of magnetic fusion energy (MFE) research on the right. Note also that the demanding power plant conditions for inertial fusion energy (IFE) actually fall outside the main figure (see chapter 11) (source: UKAEA).

the helical toroidal circulation of the plasma and less and less goes to raising the plasma temperature.

The tokamak concept is that being adopted by the now slightly renamed International Tokamak Experimental Reactor (ITER) (see chapter 10). While other magnetic confinement concepts continue to be researched, it is the tokamak concept that in both technology and policy terms seems to have the highest chance of initial success.

In section III.9.4 we consider the probable form of a first generation fusion power station based upon the tokamak concept. In figure III.9.8 we illustrate schematically the fundamental components of a fusion power plant ranging from the fusion plasma to the production of electricity.

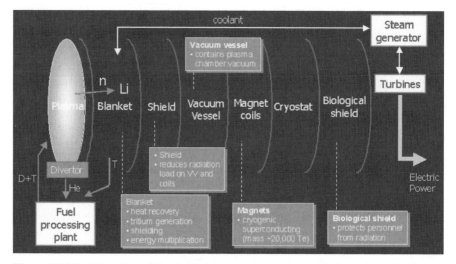

Figure III.9.8. Hierarchy of fundamental components required for the production of electricity in a tokamak fusion power plant (source: UKAEA).

III.9.2   Blankets and divertors

All good technological fields have their own special jargon and fusion is no exception. The names of the components blankets and divertors hint at their function, but for a proper appreciation of nuclear fusion energy it is important to consider these elements of a fusion power station in some detail as it is in these areas that the greatest technological challenges remain.

While plasma physics performance as given by the triple product has improved by more than five orders of magnitude in the past 40 years the performance of materials has improved only by relatively small factors. Tokamak fusion is still widely regarded as a problem in plasma physics. In fact the successes in that area have been so dramatic that the real issues of tokamak fusion are now moving towards materials science and engineering. It is not going to be possible, for instance, to raise the melting point of graphite by five orders of magnitude. Melting points and materials resilience are becoming some of the make or break issues of fusion research. These issues are fundamental to the 'fast track' approaches to fusion power production discussed in section III.9.3.

The **blanket** is fundamental to the operation of a tokamak fusion power plant. It is the first part of the outer wall of the toroidal pressure vessel (see figure III.9.8). Energy extraction in such a power plant concept relies upon the 14 MeV fast neutrons passing into the blanket and depositing their energy to the blanket in a series of collisions with the atoms of the blanket structure. The blanket must be able to cope with high-energy neutron bombardment for long periods and also with high temperatures. In addition to continuously high temperatures there are risks of localized and short-lived thermal transients that might prove damaging to blanket components.

In order to cool the blanket and to extract the thermal energy to make electricity, it is necessary to pass a fluid through the blanket structure. Conventionally this has been envisaged as being done by water/steam pipes in the blanket structure, but more recently attention has been shifting towards high-temperature gas coolants such as helium, which can drive electricity-generating turbines via a direct Brayton cycle. Figure III.9.9 shows a schematic of a modern tokamak blanket element.

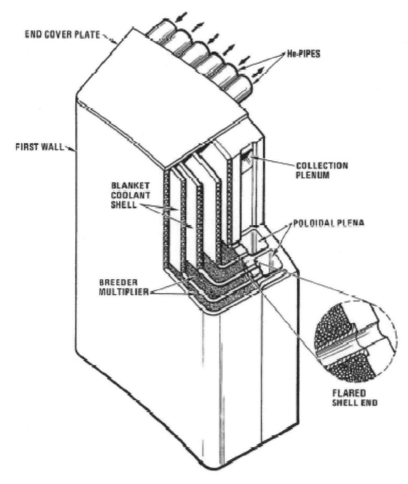

Figure III.9.9. Modern tokamak fusion blanket element showing the plasma-facing first wall, lithium-based breeder for tritium production, and helium gas cooling (source: Najmabadi [29]).

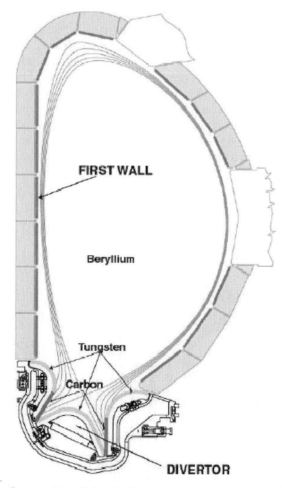

FIRST WALL

Beryllium

Tungsten

Carbon

DIVERTOR

Figure III.9.10. Cross section of the ITER pressure vessel showing the first wall and the divertor system that runs around the base of the torus. Note the crossing lines of magnetic field close to the divertor assembly. The label 'Beryllium' refers to the tokamak first wall (source: Federici *et al* [30]).

Various materials have been studied for plasma-facing components. These include graphite with its high in-plane thermal conductivity, beryllium with its low atomic mass and various heavier metals and exotic alloys (figure III.9.10). The JET experiments with beryllium turned out to be more problematic for safety than the use of radioactive tritium in the D–T plasma experiments. This is because of the high chemical toxicity of beryllium and its tendency to form dusts in plasma operations. Graphite blanket components are also of concern owing to their relative structural weaknesses, but also because of the ability of radioactive tritium atoms to intercalate between the planes of the layered graphite structure. If large amounts of graphite are used in plasma-facing components, fusion power station operators will need to be careful that they do not exceed their radioactive site limits owing to the absorption of tritium by the graphite.

For a first wall of a lightweight (low-Z) material such as graphite or beryllium, a key issue will be blanket first-wall erosion in the face of high temperatures and large fast neutron fluxes. Issues relating to heavier first-wall materials include their plasma cooling properties, whether they result in unwanted ions entering the plasma, and the need to exclude the possibility of producing problematic radioactive isotopes via transmutation of atoms within the first wall and the blanket.

The long-term stability of blanket components is a key concern because of neutron activation in D–T fusion electricity generation. The greatest

Figure III.9.11. The MASCOT remote handling equipment inside the JET vacuum vessel, being used to exchange the divertor tiles at the bottom of the JET torus (source: EURA-TOM/UKAEA).

materials challenges relate to the operation of the divertor at the bottom of the tokamak pressure vessel. It is the divertor that experiences the highest temperatures of all. The high temperatures arise as a result of transients. These plasma edge instabilities are particularly likely to occur at the divertor and can result in the sudden release of very large amounts of energy.

The development of the divertor has been a key step in the improved plasma physics performance of tokamaks in recent years. The function of this device is to remove energy and charged impurities from the plasma. In particular the helium ash arising from the D–T fusion process must be removed before it acts to cool the plasma. The concept of a divertor dates back to the early 1950s but it has only been in the past 15 years that substantial progress has been made in employing divertors successfully.

The experience at JET was that while divertors do indeed allow the tokamak to achieve higher temperatures this comes at a price of plasma stability. Gradually after much effort divertor concepts have been optimized such that higher and higher triple products can be expected. The key limiting factor, however, will be the melting point of the divertor components. Much work will need to be done in this area if reliable and sustained fusion power plant operations are to be achieved. See figure III.9.11.

III.9.3   The fast track

> *Fusion power is 50 years away, it has always been 50 years away and it will always be 50 years away.*
>
> Conventional wisdom 1958–2001

The 50 years away promise has without doubt become the single greatest problem facing the advocates of fusion. While those working on fusion rightly regard the introductory comment above as somewhat unfair and simplistic, it is also fair to say that the fusion community itself is responsible for repeating the 50 years away claim throughout the history of fusion. That this message has now been promulgated for more than 50 years means that for many policy makers the game is up for fusion—something must change. Continued calls for 50 years of financial support are now politically unacceptable.

The supporters of fusion rightly point out, however, that the statements made at all points by the research community were sincere and could have been true had it not been for the fluctuations of policy and funding. This author does not doubt the sincerity of the comments made by the fusion community; but was the falsity of these statements a consequence of technical failure, or is the matter more complex? The advocates of fusion assert the reason fusion electricity was intermittently reported to be 50 years away was not a failing of the fusion research community, but rather the waxing and waning of political and financial support for fusion over the decades, although they do acknowledge that the technical challenge was greater than originally anticipated.

When policy demands are acute, as in the mid 1970s and perhaps today, then policy makers make available significant public resources for fusion research and they expect a statement of likely timescales of return on their investment. On the basis of such funding the generation of electricity seemed to be genuinely 50 years away and statements to that effect were made. However, typically, the oil price then drops, energy security improves, the politicians lose interest in fusion, the research budgets decline and the possibility of hitting the declared 50 year target recedes.

In Europe, however, the budgets for magnetically confined fusion have recently had some resilience. This has been seen both nationally in countries such as the UK (where the last major cut was in 1990) and Germany, but also most importantly through the successive European Framework Programmes of Euratom.

The ever-receding prospect of fusion electricity has had consequences for the culture of the fusion research community. From the early days of the ZETA machine at Harwell in the 1950s, fusion research has had elements of big science. There were indeed numerous small plasma physics experiments at universities and laboratories around the world, but the community were acutely conscious of the big machines, especially JET at Culham in Oxfordshire, JT-60 at Naka in Japan and the TFTR at Princeton, New Jersey, in the United States. Despite being dedicated to the utilitarian goal of electricity production, these research efforts assumed a culture more reminiscent of the university than of industry. Large fusion research centres became reminiscent of the large particle physics research laboratories such as CERN in Switzerland or Fermilab in the United States. These laboratories are dedicated to nothing more utilitarian than understanding the fundamental properties of nature. In such an environment fusion researchers motivated themselves with the challenge of producing academic journal articles. Few fusion researchers defined career success as being the production of actual electricity using nuclear fusion. The large fusion laboratories never took on the true mission-oriented culture pioneered by NASA's Apollo programme of the 1960s and the US Manhattan Project of World War II.

It is perhaps unfair to criticize the fusion community for not having the culture of Apollo, because it is quite possible that no technical organization will ever again have such a culture. Apollo, it should be remembered, received astonishing levels of US public money through the 1960s rising to an amazing 0.75% of US GDP in 1966 [31]. That said, Apollo was distinguished not only by its vast budget but also by a sense of purpose coming directly from the President of the United States.

But why, some say, the Moon? Why choose this as our goal? And they may well ask, why climb the highest mountain? Why—35 years ago— why fly the Atlantic? Why does Rice play Texas? We choose to go to the Moon, we choose to go to the Moon in this decade and do the other things, not because they are easy, but because they are hard, because that goal will serve to organize and measure the best of our energies and skills, because that challenge is one that we are willing to accept, one we are unwilling to postpone, and one in which we intend to win, and the others too.

President John F Kennedy
speech at Rice University, Texas, 12 September 1962

Fusion never received such a clear challenge. Perhaps the time for such a challenge is now upon us.

The worldwide fusion community has in recent years attempted to break away from the paradigm that fusion power is 50 years away. The mechanism by which this is to be achieved was the result of much pressure from the British fusion community. Bringing the future timescale down below 50 years is termed the 'fast track'. Fundamental to the philosophy of the fast track is that research and development tasks that had previously been regarded as sequential should now, as far as possible, be pursued in parallel. Key to this thinking is a recognition that the ITER machine would not be able to operate as a materials test bed for plasma-facing blanket components. Such components would need to be shown to be reliable for months and years before they could be incorporated in a fusion power plant. ITER, however, will be an experimental facility. It will not operate continuously and generate sufficient long-lasting neutron fluxes to act as the main materials test bed of the fusion programme. The fast track approach requires that the materials testing required for fusion power be well under way before the completion of the ITER machine. The fast track approach therefore requires a significant effort in the medium term in the area of materials characterization under prolonged irradiation. These studies will require the construction of at least one dedicated facility based upon particle physics accelerator technology. ITER will of course contribute greatly to our under-standing of the behaviours of blanket and other plasma-facing components. For this reason ITER is designed with a modular blanket for ease of testing. For an overview of the current status of fusion materials research the reader is advised to consult the work by Muroga *et al* [32].

Interestingly a fusion-based programme of materials research with dedi-cated accelerator technologies could have synergetic benefits in other parts of the nuclear renaissance. Key constituencies with a likely interest in such developments include the research community interested in nuclear waste transmutation and in accelerator-driven systems for power production such as the 'energy amplifier' (see chapter 7) although it should be noted that fusion fast neutrons are significantly more energetic than the fast neutrons created by fission processes.

One key aspect of the fusion fast track is the elimination of one entire step in the journey from scientific research to electricity production. Original plans for that journey envisaged a sequence in which the ITER machine would be followed by a demonstration machine 'DEMO' and then a proto-type power plant known as 'PROTO'. In the new accelerated scheme the plans for the PROTO machine have been dropped in favour of a direct tran-sition from technology demonstration in DEMO to full-scale commercial power plants. As such this in turn forces the ITER research machine to be as conscious as possible of end-use considerations in order to minimize the burden facing the DEMO stage.

The fast track approach is a political imperative that can be achieved by adopting a different and in many ways more attractive technology strategy. The benefits of a fast track do not stop there, however. A fundamental benefit is the change in research culture that a fast track can generate. If the fast track can achieve fusion power in 30 years then a 22-year-old graduate enter-ing a career in fusion research and development can be optimistic that before he or she is 55 they will see the opening of a real fusion power plant. The last years of their active career would then be dedicated to the task of ensuring that fusion is an attractive and competitive choice for electricity generators planning new capital investments. Not only would such individuals see real fusion power production during their active career but also they could see its widespread adoption in the electricity industry before retirement. Such a prospect has the potential to provide a completely different motivation to the next generation of fusion scientists and engineers. In that respect the culture could be more like Apollo. A fast track approach is certainly neces-sary for such a cultural shift but it may not be sufficient. What is also needed is leadership at the highest level. As a global endeavour it is not appropriate for any one national head of state to provide such a leadership role. Perhaps the General Secretary of the United Nations or some other similar inter-national figure could be persuaded to announce the vision but, however it is achieved, the fusion community and the global public need to realize that real fusion power at some date significantly less than 50 years from now is both expected and required. Such a public challenge to the community establishes criteria for possible failure while simultaneously enhancing the likelihood of success.

III.9.4   The challenges of a fusion power station

As nuclear fusion emerges as a technology it is necessary to look ahead to how a fusion power station might actually appear. In September 1997 the European Parliament convened a meeting aiming to bring together fusion scientists and technologists with representatives from the electricity utility industry [33]. In this way it was hoped that the workshop could develop a sense of the likely realities of fusion power production. Unfortunately the organizers found it almost impossible to engage the interest of utility representatives and very few attended the meetings. This in itself reveals a powerful conclusion for the advocates of fusion. The electricity industry is now highly commercial and increasingly short-term in its planning, While the electricity industry is learning to look six months ahead, the fusion researchers were asking for industry help in looking 50 years ahead. The timescales of these two communities were so mismatched as to render attempts at dialogue impossible. Another powerful message emerged from these discussions, and it came from electricity academics. The fusion research community were advised that any fusion power plant would have to achieve a load factor of 75%, and that any fusion power plant should be operated with a need for only one refurbishment period lasting six months every ten years. As such the key message from the electricity industry experts to the fusion researchers is that any fusion power reactor must be highly reliable.

A load factor of 75% presupposes that the machine can work continuously. This is actually a non-trivial matter for a tokamak fusion power station. In section III.9.1 the tokamak is presented as a form of transformer with the plasma as the single secondary winding of the device. By analogy one might naively infer that the current running through the primary might be standard alternating current with the plasma drifting poloidally first left then right in oscillation. Such an a.c. oscillation of the plasma has been considered on paper and small experiments performed, but the associated inductances and field reversals would be so large as to be beyond the engineering limits of any large-scale machine. Rather all tokamaks operate as so-called 'd.c. transformers'. That is, the voltage applied to the primary is a single sweep from positive to negative. The resulting change in current induces the drift motion in the plasma. Unlike an a.c. transformer, the voltage applied to the primary is not reversed, the poloidal magnetic fields induced do not reverse and so the plasma drift motions are only unidirectional. This d.c. mode of operation has an important consequence: it means that a tokamak for power production would seem naturally to be a pulsed device.

One might posit that as long as the pulses could be frequent enough this would not matter because if the heat extraction involved a steam circuit, it would have sufficient heat capacity to smooth any such intermittency of the heat input. In fact, as fusion devices become larger and closer in scale to actual fusion power plants, then the transformer sweep times become longer not shorter. It is expected that with the ITER test machine a typical voltage sweep will be in the range 20–30 min; but a machine that produces power for 30 min out of every 90 min is a very odd power station indeed.

In section III.9.4 it is proposed that the thermal and electrical parts of a fusion power plant will be a conventional steam cycle. It has always been incongruous to imagine the highest technology of electricity generation systems actually producing its electricity by raising steam. It is as if 19th century engineering was being asked to produce electricity from 21st century science. The late Gordon Lake beautifully summed up the nature of this incongruity when he opined that 14 MeV neutrons is a pretty strange way to boil a kettle. It seems increasingly likely that a fusion power plant in 2040 would not employ a Rankin steam cycle in its primary cooling circuit, but much more likely primary cooling will be provided by high-temperature high-pressure helium. In this way higher thermodynamic efficiencies could be achieved and probably, by 2040, costs minimized. This is because of the benefits arising from important synergies with developments with high-temperature gas-cooled fission reactors of both a pebble bed and a prismatic design (see chapter 6). Whether fusion power stations will actually employ a secondary steam cycle or directly drive high-temperature gas turbines with the hot helium in a direct Brayton cycle remains to be seen. Such a move away from indirect steam cycles and towards direct helium cycles would, however, further increase the need for any fusion power plant to operate in steady-state mode.

It is important to stress that the obstacles to steady-state operation are not primarily a consequence of the properties of the plasma. Rather the difficulties with steady-state operation are related to the magnetic, electrical and mechanical properties of the tokamak machine itself. While the machine is operated as a d.c. transformer it is on first impression inevitable that some form of pulsing will be required. It is widely believed that plasma stability will continue to improve such that the plasma itself could be maintained almost indefinitely. It is the probable need for the plasma to be driven magnetically that forces planners to confront the issues of pulsed operations. It should be noted that in recent years it has become clear that there is the possibility of self-generated circulation in a tokamak as a result primarily of a phenomenon known as the 'bootstrap current', whether this will be sufficient to allow for continuous tokamak operations is not yet clear.

Another fundamental issue for fusion power is its relationship to the evolving nature of the electricity industry. It seems inevitable that if nuclear fusion is to be used to generate electricity; then the power produced will be base load in the best traditions of large-scale centralized generation. In part this is a consequence of the GW power output levels, but also it is a consequence of the need for several hundred MW of power in order to black start a fusion power station. For reasons of both power needs and generation capacity, fusion power will require reliable and extensive high-voltage electricity grids. Given the several decades of further work before fusion power plants become a commercial reality there is a need to monitor the likely relationship between fusion technology and the needs of the electricity industry. A large-scale move to distributed generation, as posited in Walt Patterson's book *Transforming Electricity* [34], would mean a significant diminution in the need for fusion power. As earlier discussion of the pebble bed modular reactor makes clear, nuclear power is not inevitably a large-scale base load technology, but it does seem that magnetic confined fusion power would be. Such considerations motivate consideration of fusion as a heat source for thermochemical hydrogen production as discussed in section I.3.4. By such means fusion energy might achieve a prosperous future despite any moves away from large scale base-load electrical generation.

Chapter 10

Magnetic confinement fusion—future technologies

It noteworthy that all major candidate geometries for magnetic confinement fusion rely on some form of toroidal vacuum vessel. One might argue that the Z-pinch approach discussed in chapter 11 is an example of non-toroidal magnetic confinement, but equally one can take the view that the physics of the Z-pinch owes more to the inertial confinement paradigm. Hence in this text the Z-pinch is included in chapter 11.

The use of a torus to magnetically confine a plasma goes right back to the ZETA experiment at Harwell in Oxfordshire in the 1950s. The over-hyped ZETA machine was a toroidal pinch with several similarities to the later tokamaks. Toroidal pinches rely on a transformer action with the toroidal charged plasma being the secondary winding. However, unlike a tokamak only weak toroidal fields are applied from external coils. The toroidal pinch relies on very large induced plasma current giving rise to a self-induced pinching of the plasma that brings it away from the walls of the pressure vessel. The plasma is less actively controlled than in the tokamak concept, and as such the simplicity and elegance of the approach is appealing. Unfortunately the simple toroidal pinch did not survive the 1960s as a serious contender for fusion power generation because in every case there was found to be a rapid loss of energy from the plasmas produced, i.e. only very short confinement times could be obtained.

Towards the end of the ZETA investigations an unexpected phenomenon was observed: the toroidal magnetic field was found to reverse itself spontaneously inside the pressure vessel towards the edge of the plasma. The phenomenon of the 'reverse field pinch', as it became known, added toroidal plasma stability and was to prove useful in improving tokamak and stellarator stability. These approaches had already emerged as front-runners in magnetically confined fusion research when field reversal was first observed and understood.

ZETA comes from the early days of optimism for fusion energy production. It is not the purpose of this book, however, to survey the technological and scientific history of magnetically confined fusion. C M Braams and P E Stott have already provided an extensive and illuminating history of such matters [5]. Rather, we aim here to look to the future and in this respect the dominant issue is ITER.

III.10.1   ITER

In the global quest for sustainable fusion power one project stands above all others. Originally known as the International Thermonuclear Experimental Reactor the name transformed to the International Tokamak Experimental Reactor, or simply ITER. The fusion community now remind us that 'iter' means 'the way' in Latin. In part the shift from acronym to name is a somewhat unsuccessful attempt to avoid confusion between the current reduced-scale machine and the original more ambitious proposals for the International Thermonuclear Experimental Reactor.

In early 2004 ITER is a truly global collaboration. Participating countries include the member states of the European Union, Japan, the United States, South Korea, China and Russia. The design has been reduced in scope and ambition from the overpriced plans of the 1990s. The current ambition is for a US$10 billion project (total lifetime cost) producing a burning plasma of at least $Q \approx 5$ (probably around $Q = 10$) by around 2015. The triple product for the ITER device is expected to be around 3×10^{21} keV m^{-3}s. See figure III.10.1.

The former ambitions for an ignited plasma ($Q = \infty$) have been dropped as part of the cost reductions needed after the original proposal, but equally it has been realized that ignition is not actually desirable. With its new design parameters it is planned for a fusion power of approximately 500 MW produced from a machine only slightly smaller than is expected for a commercial fusion power plant. The thermal power of a fusion power plant is expected to be eight times larger at approximately 4 GW [35]. The ITER machine will be the first large-scale fusion experiment investigating D–T plasma behaviour under reactor-like conditions. To do this the ITER machine will use some of the largest superconducting magnetic coils ever developed.

In July 2004 the location of the ITER machine still remains undecided. Two candidate sites are competing for the right to host the project. The European bid centres on Cadarache in Provence, southern France, while a Japanese bid proposes construction at Rokkasho-Mura at the northern end of Japan's largest island, Honshu. At the point at which it became clear that Canada's preferred site in Clarington, Ontario, would not be adopted, the government of Canada chose to withdraw from the ITER project entirely.

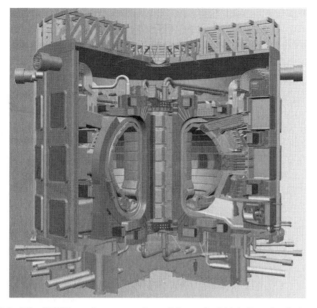

Figure III.10.1. The ITER project as envisaged in early 2004. Despite being scaled back from the original ambitions of the 1990s the ITER machine will still be extremely large and complex. Note small figure to the bottom right, which gives a sense of the scale of the machine (source: ITER Project).

The cost of ITER is a complex thing to define given the highly international nature of the project and the fact that only a fraction of the contributions will be supplied in the form of cash. Figure III.10.2 illustrates the nature of the various contributions expected. The ITER

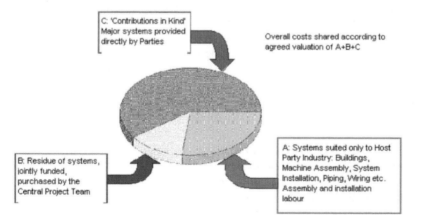

Figure III.10.2. Expected basis of ITER procurement and financing (source: ITER Project).

project management team divides the items to be purchased into three categories:

A.   Those that can only sensibly be purchased by the host country.

B.   Those which are of minor technical interest or size, and whose cost burden must therefore be shared by all Parties. For these items a centrally administered fund can be established.

C.   Items of interest to all the Parties due primarily to their high technology content. To ensure each Party obtains its fair share of these items, the Parties must agree beforehand which ones each will contribute. To do this, all Parties must agree on their value to the project. This requires an agreed valuation of items. Each Party will contribute its agreed items 'in kind', using the purchasing procedures and funding arrangements it prefers. Thus the actual costs to each Party may not correspond to the project valuation—it may differ due to competitive tendering as well as different unit costs.

(Source: http://www.iter.org)

This level of care is required because of the need to ensure a reasonable level of *juste retour* to the contributing countries. That is, individual countries should receive benefits consistent with their level of contribution. Each country contributes to the project for reasons of utilitarian self-interest. Part of the self-interest relates to the ultimate goal of environmentally benign, economic and reliable electricity generation, but a greater concern is to ensure that the industrial and research base of each contributing nation should benefit to an appropriate extent from the involvement of that country in the project. The greatest concern is to ensure that the host country contributes an appropriate amount given that it is likely to benefit to the greatest extent from the innovations and knowledge generated during construction and operations and from the general economic boost of hosting such a large facility on its territory. Countries naturally want to maximize the extent to which they contribute in kind (category C) and minimize the extent to which they must contribute cash (category B).

The relative proportions of equipment and infrastructure required are shown in figure III.10.3.

It is fundamental to the strategy of tokamak-based fusion power research that the ITER machine is in no way a power station. Originally the path from ITER to commercial power generation had four steps: the ITER research machine, a demonstrator known as DEMO, a prototype reactor known as PROTO and then a fleet of commercial power plants. In the move towards fast track approaches the PROTO and DEMO stages have been merged, but the ITER step remains largely unaffected (see

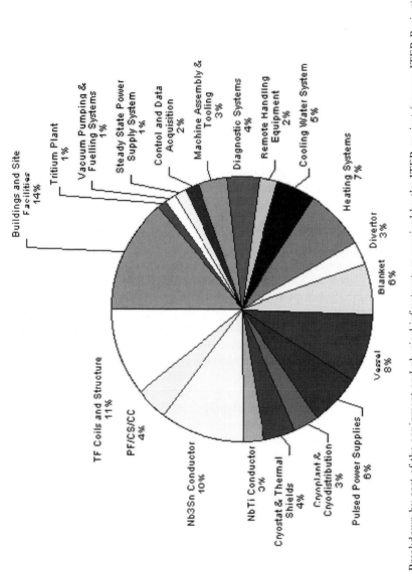

Figure III.10.3. Breakdown by cost of the equipment and physical infrastructure required by the ITER project (source: ITER Project).

section III.9.3). When defending the position that the ITER team should avoid any ambitions to generate electrical power, mention is sometimes made of the difficulties of the Advanced Gas-cooled fission Reactor programme in the UK in the 1970s. It can be argued that the AGR programme revealed the dangers inherent in attempting simultaneously to roll out a fleet of new power plants while also seeking to optimize the design. Any attempt to mix the launch of a new type of power plant with fundamental research into the viability of such a plant is probably too ambitious and likely to result in disappointment. The damage that would result from such a high-profile failure would be fatal to ambitions to make fusion power a serious and viable technology. The counter argument runs that the patience of policy-makers and the public should not be regarded as inexhaustible and that there is a need to celebrate something useful emerging from the fusion programme at the earliest possible moment. A small experimental steam or high-temperature helium system to extract heat from the blanket and to drive a token turbine sufficient to produce the first fusion electricity, say to boil a kettle with fusion power on the TV news, could have enormous public relations benefits to the fusion project. The pressures towards a fast track for fusion relate to the need for rapid and visible outputs. The changes to fusion strategy are not, however, restricted to an accelerated materials science programme in the years leading up to ITER commissioning. Substantial changes in strategy have also been adopted for the years after ITER.

While ITER is most definitely not a power station, it is also not a radical leap into the dark. In a way reminiscent of the Apollo space programme of the 1960s, the ITER project starts from the premise that each of the key technological elements has been demonstrated. What is required is to combine these elements (the superconducting magnets, the high-temperature materials for the blanket and the divertor, the D–T plasma itself) for the first time in a reactor-like configuration. As with Apollo, the fundamental challenges are not technical but organizational. Just as the commercialization of space has revealed examples of cultural mismatch and cognitive dissonance between the government agencies and private industry, so it is likely that the organizational structures developed by the fusion community to deliver ITER will have difficulties dealing with the power utilities and engineering consortia that might be expected to be the developers of a fleet of commercial fusion power plants around the middle of the century.

One possibility might be to engage the major oil companies in support for fusion development. These companies are sufficiently large as to be able to engage with highly speculative and expensive ventures. On the time-scale of fusion's development these companies will need, as fossil fuel resources become depleted, to complete the shift from being oil companies to being energy services companies. The timescales of oil depletion are coincidentally expected to match those of significant global climate change.

This can only increase the pressure on the oil majors to innovate and move forward (see the discussion of nuclear hydrogen in section I.3.4).

III.10.2 Spherical tokamaks

In this chapter we look to future technologies in magnetically confined fusion. ITER, however, is by no means the only idea in circulation.

The ITER machine will be at the big end of big science. Its culture is one of a large multinational team spending large budgets on a long-term complex project. Developments of this type are fundamental to technological innovation and have led to developments such as radar, nuclear weapons and manned space flight. Fusion, however, is also receptive to developments from poorly funded small teams with radical ideas, and probably the best illustration of this paradigm has been the spherical tokamak.

The Discovery Channel has produced an excellent television documentary on this subject under the title *Bottling the Sun*. The story is so rich and inspirational that it could even sustain dramatization.

The story starts with small experiments in Australia, Germany and the United States. These experiments were looking at unusual geometries for magnetic confinement fusion with names like the 'spheromak' and the 'rotamak'. In the late 1980s at Oak Ridge National Laboratory in Tennessee, USA, Martin Peng proposed the construction of a spherical torus experiment to test his 1985 theory that a tokamak shaped like a cored apple rather than a ring doughnut would suffer less plasma turbulence and be able to retain its energy longer. Peng's attempts to secure funding in the United States were, however, unsuccessful.

On the other side of the Atlantic the fusion group at Culham in Oxfordshire were intrigued by Peng's idea. The laboratory director R S (Bas) Pease asked his staff to look into the merits of the idea and plasma theorist Alan Sykes took the lead. At this point the Culham team took a bold and imaginative decision. They would not follow the orthodox path that had frustrated Martin Peng. They would not spend months of effort pursuing government funding, they would simply get on with implementing the idea without any real resources at all. In any dramatization of this story a central element must be the team meetings held in the appropriately named Machine Man Inn in the quaint and pretty English village of Long Wittenham. The idea that crucial design ideas for the new form of tokamak were developed on the back of beer mats is not too far from reality.

The meetings were held at lunchtime as few of the laboratory's scientists could devote time officially to the project. Also very little money was available. These constraints forced the team to raid the Care and Custody Store of Culham Laboratory. This store was basically a junk room storing of bits of old experiments, some of which dated back decades.

As the project developed from 1987 onwards it took on the name START (Spherical Tight Aspect Ratio Tokamak). One of the key innovations was the realization that the shape of the main pressure vessel did not need to follow the shape of the plasma itself. This allowed the team to drag into the laboratory a decades-old cylindrical aluminium pressure vessel that had been lying out in the wind and rain for years. With liberal use of high vacuum sealant, the vessel was pumped out, and an acceptable base pressure was achieved despite significant outgassing from the aluminium. While aluminium is a far from ideal material for high-vacuum work the solution adopted was good enough, quick enough and cheap enough.

By 1990 research was under way and in 1991 it was demonstrated that plasma could indeed be confined in a spherical geometry. From this point on some seriously expensive pieces of equipment were attached to the experiment. These included in 1994 the loan from Oak Ridge National Laboratory of a neutral beam injector for plasma heating.

In the late 1990s a series of remarkably experiments were undertaken on START culminating with observations in 1998 of a world record value for the plasma pressure ratio β (see section III.9.1). The record set by START beat the previous record by a factor of three. See figure III.10.4.

In many areas of science and technology from aircraft to bridges the beauty of a design is an accurate reflection of technical merit. The START experiments of the late 1990s gave the world some of the most beautiful images from any field of science and technology. Figure III.10.5 shows such a photograph.

START was followed in 1999 by developments back in the birthplace of the tokamak concept—Russia. The Globus-M tokamak in St Petersburg has been developed for a series of experiments to study the behaviour of spherical plasmas [36]. The Globus-M team have been devoting much effort into neutral beam injectors at the megawatt level. By 2004 this has been joined by a series of spherical tokamak experiments on four continents.

In recent years spherical tokamak research has entered the phase of large-scale projects with major new facilities being constructed at two of the historical homes for big magnetic confined fusion experiments: Culham in Oxfordshire, England, and Princeton in New Jersey, USA.

At Culham a proof of principle machine has been constructed under the name of the Mega-Amp Spherical Tokamak (MAST) [37], while in the United States another mega-Amp machine has been built—the National Spherical Tokamak Experiment (NSTX) [38].

These machines differ from START in that they both return to the paradigm of large government projects with concomitantly large budgets. Both MAST and NSTX were made possible by the demonstration on START that indeed spherical tokamak geometries do show higher degrees of stability, that the plasma does adopt the required shape, pulling away from the walls of the pressure vessel, and most importantly that the geometry

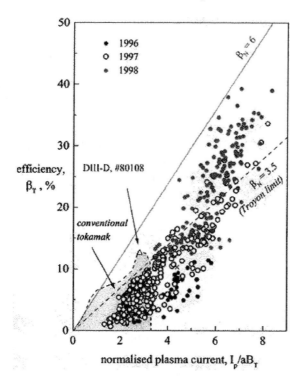

Figure III.10.4. A world record $\beta = 40\%$ was achieved on the START spherical tokamak at Culham in Oxfordshire in 1998 (source: UKAEA).

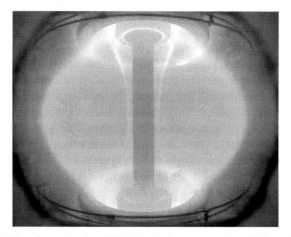

Figure III.10.5. Hot plasma in the START spherical tokamak at Culham in Oxfordshire UK (source: UKAEA Culham).

allows for the highest plasma temperatures and pressure ratio β. These attributes out-perform the behaviour of more conventional tokamaks such as JET and ITER and have led to questions being raised as to whether the ITER path is indeed the right one.

The reality for nuclear renaissance in the fusion sector, however, is that spherical tokamaks are at least a decade behind their conventional tokamak cousins. For instance, no spherical tokamak is capable of performing radioactive D–T experiments of the type performed by JET and TFTR in the 1990s. In addition the magnetic coils in the spherical tokamaks are placed relatively close to the plasma—in the case of START and MAST, actually inside the pressure vessel. As such, innovations are required because in D–T plasma fusion the 14 MeV neutrons produced would be very damaging to the magnetic field coils unless they could be shielded and protected properly. The dominant concern is actually the centre column of a spherical tokamak reactor. It is expected to require regular replacement given the irradiation that it will receive and the fact that it will not be easily shielded owing to the highly constrained geometry of these devices.

The promise of the spherical tokamak is so great that plans have been advanced for a strategy to go from today's research machines to a fully-fledged power station. In this scheme the MAST and NSTX would be followed by a spherical tokamak capable of operating with D–T plasma, to be known as the Next Step Spherical Torus (NSST) [39]. Thermal fusion power creation of up to 60 MW is expected for this machine. The NSST would be followed by an intense programme of materials testing (informed by experience gained in the ITER project) and the development of a prototype steady-state spherical torus reactor known as the Component Test Facility or CTF [40]. The CTF machine could then be followed by a full-scale demonstration power plant known as 'DEMO'. This would have steady-state thermal fusion power of approximately 3 GW and a plasma major radius of only 3.4 m. The CTF machine will also be of great interest to conventional tokamak developers as they follow the ITER-based trajectory. Components of relevance to ITER and its successors could be usefully tested on the CTF. It is possible that spherical tokamaks might never be suitable as the basis of an electricity generating power plant. It is also possible that the development of the spherical tokamak might end with the CTF. Even in such a situation the boost given to the ITER programme by results and insights from the CTF would probably be sufficient to justify the resources devoted to spherical tokamak development. In fact the potential of the spherical tokamak is greater than those modest ambitions. It has a real potential to be a key technology in the history of fusion energy and the nuclear renaissance.

Separately the United States has pulled together a wide-ranging collaboration under the name of the 'ARIES Team' to look innovatively at issues in fusion energy. As part of its work, the team was asked by the

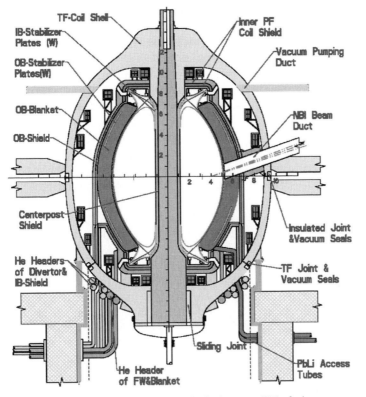

Figure III.10.6. Elevation view of 1 GW spherical torus (ST) fusion power core as developed by the ARIES-ST study as part of a conceptual design for a future ST power plant (reprinted from Galambos J D, Strickler D J and Peng Y-K M 1999 Systems cost and performance analysis for a spherical torus-based volume neutron source *Fusion Engineering and Design* **45**(3) 335–350, with permission from Elsevier).

US government to estimate the key attributes of a potential spherical torus fusion power station. The ARIES-ST study considered a machine of the type shown in figure III.10.6.

The ARIES-ST concept envisages a 1 GW thermal plasma power machine with a plasma major radius of 3.2 m. The team estimates a total capital cost of approximately $4.5 billion in 1992 US dollars. Fundamental to the ARIES-ST concept is the efficient maintenance and serviceability of the reactor. This is enhanced via plans for interchangeable fusion power cores. The fusion power core is illustrated in figure III.10.6. Figure III.10.7 illustrates the integration of the interchangeable power core with a movable and reusable support platform. Heat is extracted from the reactor using two systems. One is reliant on helium gas (cf. chapter 6) with the other utilizing lead–bismuth eutectic liquid metal technology (cf. chapter 7).

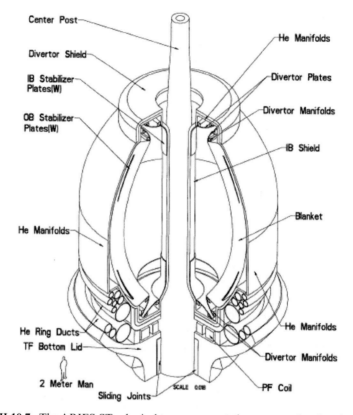

Center Post
Divertor Shield
IB Stabilizer Plates(W)
OB Stabilizer Plates(W)
He Manifolds
He Ring Ducts
TF Bottom Lid
2 Meter Man
Sliding Joints

He Manifolds
Divertor Plates
Divertor Manifolds
IB Shield
Blanket
He Manifolds
Divertor Manifolds
SCALE 0.08
PF Coil

Figure III.10.7. The ARIES-ST spherical torus power station concept showing the blanket, divertor manifolds and the helium heat extraction system. In the ARIES-ST concept the entire power core is removable and replaceable (reprinted from Galambos J D, Strickler D J and Peng Y-K M 1999 Systems cost and performance analysis for a spherical torus-based volume neutron source *Fusion Engineering and Design* **45**(3) 335–350, with permission from Elsevier).

All fusion power plant concepts face the task of demonstrating reliable operations. This is followed by the difficult task of demonstrating economic viability. Despite these difficult hurdles it seems as if spherical tokamaks do indeed have the possibility of emerging as a technology for the nuclear renaissance.

The magnetic confinement fusion research community have difficulty balancing the promise of the embryonic technology of spherical tokamaks with the momentum and inertia that is needed to push forward more conventional approaches based upon the ITER type of geometry. In fact, given the head start achieved by the ITER paradigm, both should be pursued with vigour. A useful insight comes from comparing the current status of spherical tokamaks with that of the aviation jet engine in the early 1930s.

The promise of the technology and its superior attributes had been demonstrated in the laboratory but the technology was still at least a decade away from being applied in real aircraft. Simultaneous with the jet research efforts of the 1930s the civil aviation industry grew using radial internal combustion engine technology. By the time that jet engines became suitable for civil aviation in the 1950s the industry had already developed a market and a functioning infrastructure. This analogy gives a clue to a possible future of fusion power in a nuclear renaissance—development of industrial-scale activity based upon the ITER plan, followed by a possible later move to related, but superior, concepts. The spherical tokamak is one such candidate, but it is not the only one. Another is a concept known as the stellarator.

III.10.3   Stellarators

The stellarator concept is one of the oldest in fusion research, having been developed by Lyman Spitzer of Princeton University in the early 1950s [42]. The key to the stellarator concept is that the helical toroidal motion of the plasma ions and electrons is not driven by a transformer action but rather is provided in its entirety by the use of externally applied fields. In order to do this the field coils themselves must generate a rotational transform, and need therefore to be geometrically far more complex than those used in tokamaks. The windings of a stellarator ensure closed magnetic surfaces and hence low plasma losses. With no requirement for transformer action and an applied current, stellarators naturally operate in continuous steady-state mode. This has long been regarded as a key advantage of the concept for real power station applications. However, the lack of a transformer action eliminates the possibility of resistive heating within the stellarator plasma and until recently it seemed unlikely that sufficient temperatures could be achieved for efficient fusion to take place. Thankfully there have been significant recent advances in plasma heating technology via microwaves and neutral beam injection, Previous attempts to introduce energy for plasma heating had relied on pulses of induced current, but these caused kink instabilities, eroding one of the key benefits of the technology [43]. The real innovation, however, that has made the stellarator a truly practical proposition has been the improvement in computer performance. Only in recent years has it been possible to model the performance of complex stellerator magnet designs on a computer.

One of the distinctive features of stellarators is the obvious complexity of their construction (see figure III.10.8). The world's leading stellarator project, the Wendelstein 7-X, is under construction at Greifswald in eastern Germany. The project team has tackled the complexities of design and manufacture through a modularized approach.

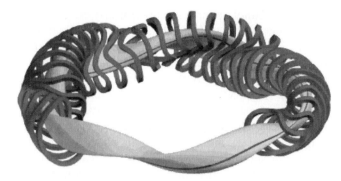

Figure III.10.8. The conceptual arrangement of magnetic field coils in the Wendelstein 7-X stellarator under construction in Greifswald, Germany. The field coils enclose a plasma with induced helicity (source: Max-Planck Institut).

On completion in around 2010 the Wendelstein 7-X machine will achieve performance levels not far behind those reached by the big tokamaks (JET and TFTR) in about 1990. As such, stellarators, while scientifically interesting, remain far behind orthodox (ITER-based) approaches to magnetically confined fusion. It seems likely that while they will continue to give insight into magnetic confinement physics and to continuous operational modes, they will need many years if they are to overtake the dominant tokamak approach to be come a major technology of the nuclear renaissance.

Chapter 11

Inertial confinement fusion

Chapters 9 and 10 discussed the dominant paradigm in attempts to produce electricity from nuclear fusion—magnetic confinement fusion. In this paradigm a hot low-density plasma is held within a doughnut-shaped torus by means of magnetic fields. In such research machines the plasma is sustained for several seconds. In an actual power station the plasma would last many hours and might even be maintained continuously.

Experience from the development of thermonuclear weapons makes clear that magnetic confinement approaches are not the only route to releasing nuclear fusion energy. A thermonuclear weapon or hydrogen bomb has two fundamental parts known as the primary and the secondary. The primary consists of a fission bomb (atomic bomb). On detonation this produces a massive radiation burst, which is focused on to a fusion pellet comprising the secondary part of the weapon. Insights gained by weapons researchers into the fundamental principle underlying the operation of the thermonuclear secondaries could provide an alternative route to fusion electricity production completely independent of the magnetic confinement approach. The more weapons-related approaches based upon the forced compression of a dense plasma pellet are known as inertial confinement fusion (ICF). Given the intense pressures involved, none of the routes to ICF-generated electricity would be based upon a continuous fusion process. ICF devices are inevitably pulsed devices. Each pulse of energy would arise from the compression and fusion of a very small fuel pellet. Typically each fusion pulse might last only a few nanoseconds (billionths of a second) [44]. Naturally the pellets used in an ICF device at only a few millimetres across would be much smaller than those used in the secondaries of thermonuclear weapons, and of course the energy yields from the fusion experiments would be enormously lower than in any type of nuclear weapon. It is anticipated that an ICF-based power station would operate at approximately five fusion pulses a second.

Just as magnetic confinement fusion has several distinct approaches (the conventional tokamak, the spherical tokamak, the stellarator etc.), so too

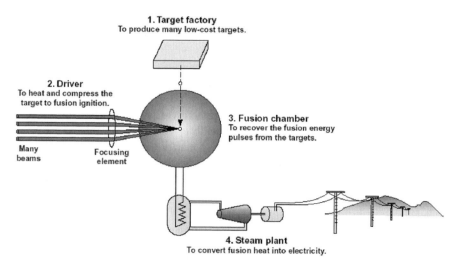

1. Target factory
To produce many low-cost targets.

2. Driver
To heat and compress the
target to fusion ignition.

3. Fusion chamber
To recover the fusion energy
pulses from the targets.

Many
beams

Focusing
element

4. Steam plant
To convert fusion heat into electricity.

Figure III.11.1. Fundamentals of a conceptual inertial confinement fusion power station
(source: Lawrence Livermore National Laboratory [45]).

does ICF (figure III.11.1). The most developed ICF concept is the direct use
of symmetrically converging laser beams to compress a fusion pellet. An
alternative that has made rapid progress in recent years is the indirect use
of x-rays to compress the plasma in a 'Z-pinch'. Also, work is under way
to use ion particle beams to compress fusion pellets. Each of these
approaches will be discussed in this chapter.

While the electricity demands from magnetic confinement fusion experiments are large, the peak power levels demanded for inertial confinement
fusion experiments are positively breathtaking. For instance the Z-machine
at Sandia National Laboratory (a Z-pinch experiment) requires a peak
electrical power of 50 TW (approximate pulse 20 MA over 100 ns). The Z-
machine therefore accumulates sufficient power to be able discharge roughly
50 times the power in the entire US electricity system over a small fraction of
a second (the shot) [44]. The electrical power levels of the National Ignition
Facility in Livermore, California, will be even more impressive—requiring
ten times more peak power than the Z-machine. The NIF project's
500 TW will also be required for only a few nanoseconds, but in that short
period the machine will be using more electrical power than the rest of the
world put together [45]. The need to accumulate such power levels and the
very short one-shot nature of the ICF process ensure that despite the likelihood that the NIF inertial confinement machine will achieve ignition before
any tokamak, it is the magnetic confinement approach to fusion that will
likely yield a practical power station most quickly.

Common to all ICF experiments is the use of a fusion pellet surrounded
by an ablator [44]. It is the pellet that contains the material to be fused,

releasing nuclear energy according to the principles discussed in chapter 9. As with magnetic confinement, it is expected that the first fusion reaction to be developed will be that between deuterium and tritium—the D–T reaction discussed in chapter 9. In the case of magnetic confinement the deuterium and the tritium are introduced as gases and they form a low-density plasma. In ICF experiments the deuterium and tritium are prepared in a cryogenically frozen pellet of solid material. The function of the ablator is to aid the compression of the fusion pellet. In the ICF approach the incoming radiation (laser light, x-rays or particle beams) heats the outer surface of the ablator. This causes atoms at the surface to boil off so energetically that they act like a rocket engine pushing down on the fusion pellet. Unlike a rocket, however, where the intention is to use the resulting pressure to cause movement in a single direction, in this case the requirement is for the rocket pressure to be absolutely equal on the pellet from all directions and for the centre of mass of the pellet and its symmetric shape to be unchanged—simply for it to be compressed exactly where it is. Possible ablator materials include plastic and beryllium, although beryllium doped with other metals such as copper may well work best of all [46]. One of the key problems of ICF is to ensure uniform compression and for this a uniform and smooth ablator surface is essential. Plastic ablators are most easily manufactured and they have the benefit of being transparent so that researchers can see that they have been properly filled with D–T fuel. Unfortunately plastic ablators do not function nearly as well as would be liked. Beryllium is a much better ablation material but it is a light grey metal and as such completely opaque. It is therefore not possible to verify that all is well with the D–T fuel held inside. Furthermore, making a perfectly smooth and symmetric small beryllium sphere is far from straightforward. Beryllium is a brittle metal that is particularly difficult to machine, not only because of its unusual metallurgy but more importantly because of the severe health hazards posed if its dust were to be inhaled, eaten or absorbed into a wound. Very few facilities in the world are equipped to machine beryllium components.

As regards the contents of the ablator, researchers can be more confident that the D–T fuel itself will have a smooth surface. This is a consequence of the radioactivity of tritium and the fact that surfaces typically have lower melting points than bulk material. The radioactivity causes sufficient heating to melt the surface of the frozen pellet. The melted surface layer then smoothes out imperfections such as bumps and scratches, giving the required smoothness of 50 nm on a millimetre-scaled object [46].

A key difficulty common to all inertial confinement approaches to nuclear fusion is Rayleigh–Taylor instability. This is a particular class of instability for imploding plasmas, which affects the behaviour of the D–T plasma in all ICF experiments and the behaviour of x-ray producing metal ion plasmas in Z-pinch experiments. Basically, if there is the slightest fluctuation or unevenness in the surface of imploding plasmas then it tends to grow and to cause the plasma to squirt out in all sorts of directions. Jeremy Chittenden draws the analogy of a glass of water turned quickly upside down [44]. A naive physics-based view would say that the water will stay in the glass for seconds because a vacuum cannot form behind the water as it tries to fall under gravity. This is the same as the naive view that a uniform sphere of plasma subjected to a uniform surface pressure will compress to smaller and smaller spheres quite smoothly. Just as surface instabilities cause the water to fall out of the glass almost immediately, so Rayleigh–Taylor instabilities cause the imploding plasma to become mis-shapen and useless. The only hope is to compress the plasma as uniformly and, even more importantly, as quickly as possible.

The key to inertial confinement fusion is the need to deposit pressure and energy on the outer surface of the ablator in a uniform, symmetric and sudden fashion. There are three main ways in which this may be achieved and these are discussed in the following three sections.

III.11.1   Laser-driven inertial confinement fusion

In chapter 9 we considered the history of civilian research programmes dedicated to eventual goal of electricity generation using nuclear fusion. The dominant approach, magnetic confinement in toroidal vacuum vessels, has long suffered from the problem that its useful outputs remain decades away. Inertial confined fusion (ICF) research, however, has a quite different heritage. If anything the idea of using ICF techniques to generate electricity is a spin-off from the primary goal of the research: improved thermonuclear weapons and a better understanding of their long-term reliability. Perhaps the clearest example of this paradigm is the National Ignition Facility (NIF) (figure III.11.2) under construction at Lawrence Livermore National Laboratory in Livermore, California. This astonishingly ambitious machine is dedicated to the stockpile stewardship of the United States thermonuclear weapons inventory in the context of a US policy against actual weapons tests. As mentioned previously, thermonuclear weapons are two-stage devices and it is the ambition of the United States to be able to test the physics of the fusion-powered secondaries of these weapons without the need for the energy from the fission-based primaries of thermonuclear weapons. The stated ambition of the NIF is captured in its very title—the 'ignition' of a fusion plasma. To do this the NIF will bring together 192 separate ultraviolet beams from the world's most powerful laser. These will be brought together on the surface of a beautifully crafted ultra-symmetric, ultra-smooth gold hohlraum (see the box *The Hohlraum* on page 294). Heated in nanoseconds to millions of degrees, the hohlraum will radiate x-rays symmetrically down upon a D–T fusion pellet surrounded by a copper-doped beryllium ablator. The laser required to do this is enormous. It struggles to fit into the football

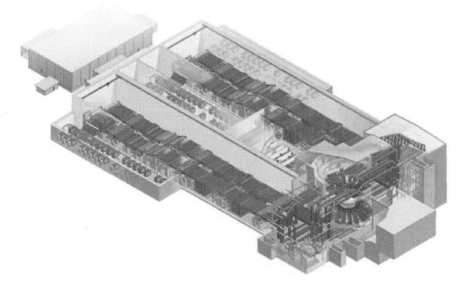

Figure III.11.2. Schematic view of the National Ignition Facility under construction at Lawrence Livermore National Laboratory in California. To the right is the NIF target chamber (source: Lawrence Livermore National Laboratory [47]).

pitch-sized building specially constructed to accommodate it. The aim is for the NIF laser to deposit 1.8 MJ of energy on to the hohlraum in only a few nanoseconds.

From the outset the NIF project has been beset by difficulties at all levels from the purely technical to the political and the administrative. Most bizarrely, almost immediately after the launch of the project the first Director resigned ater it was revealed that he had never finished the PhD at Princeton University referred to in his job application. A separate problem was that the NIF senior management did not share with the US Energy Secretary Bill Richardson the full extent of the problems they were encountering. Their overly optimistic advice caused him in turn to give unduly upbeat public assessments of the progress of the project [48].

Since the project was first approved by the US Congress in 1993 the budget has skyrocketed from original estimates of $2 billion. By August 2000 the US General Accounting Office was already estimating a total cost of more than $2.8 billion. Since those estimates fiscal pressure has eased somewhat with significantly increased US budgets for nuclear stockpile stewardship, and national security generally. The original commissioning date of 2002 has been missed and operations are not expected before 2008 at the earliest. Originally it was planned that the NIF would fire six shots per day, but by 2000 this ambition had already been cut back to just two shots per day. Researchers in high-power laser physics see the culture of

The Hohlraum

Inertial confinement fusion (ICF) falls into two direct types the direct and the indirect. Direct ICF relies upon the symmetric convergence of ion beams or laser light to compress directly a D–T fuel capsule. Indirect fusion uses a symmetric converging pulse of x-rays to achieve the same goal. These x-rays are emitted by a structure known as a 'hohlraum'. Inertial fusion energy can make use of hohlraums in laser-driven systems and in static Z-pinch devices.

Hohlraum is a German word meaning cavity. In radiation thermodynamics it has a particular meaning. It is a cavity in which the walls are in radiative equilibrium with the radiant energy within the cavity. As such, the hohlraum relates to concepts of basic physics, as it is a textbook source of blackbody or Planck radiation. It is a perfectly absorbing and radiating body whose radiative properties are given by its temperature. A hohlraum is always a highly opaque enclosure typically made of a heavy metal such as gold or lead. When the walls of the cavity are energized, for instance by laser light or the x-rays emitted by a Z-pinch metal plasma, they emit a highly even flux of x-rays sufficient to drive the compression process of a fusion capsule held within the hohlraum.

The x-rays from the hohlraum heat the ablator around the fusion capsule. The hohlraum x-rays and the vaporizing ablator compress the frozen D–T fusion fuel, sending shock waves through it. One part of the fusion fuel will first achieve the temperature and pressure conditions needed for significant fusion to occur and a 'hot spot' is said to form. This can then in turn ignite the entire fusion fuel pellet, creating a burst of inertial fusion energy.

their discipline changing before their eyes. In the 1990s graduate students could be placed in charge of shots at some of the world's largest lasers such as the VULCAN machine at the Rutherford Appleton Laboratory in England. Once the NIF is operational each shot will be more reminiscent of the launch of a probe into space. Huge teams of experts will gather together under the leadership of experienced senior scientists anxiously watching computer monitors in moments of collective anticipation. With shots on NIF worth more than a million dollars each, no graduate student is going to be allowed to run the laser through the night-time hours.

In every way the NIF is an ambitious undertaking and in some technical ways it has already come close to being over-ambitious. First the NIF team had to confront the safety issues of storing megajoules of electrical energy in

Figure III.11.3. The NIF target chamber is readied for its unveiling ceremony on 11 June 1999 (source: Lawrence Livermore National Laboratory [49]).

advance of a shot. Each of the 200 ultra-large capacitors is designed to be safe in the event of explosive failure.

The NIF laser system uses enormous potassium dihydrogen phosphate (KDP) crystals to triple the laser frequency from the visible region into the ultraviolet. This dramatically reduces problematic heat generation in the laser, but at the price of far greater damage to the optics. In 2000 it was estimated that the laser optics on NIF would need to be replaced after only 50–100 shots [46].

Whether the NIF will ever achieve 'ignition' (see chapter 9) is far from certain. Whether it will ever add to the United States' understanding of its thermonuclear weapons is also far from clear. What is clear, however, is that NIF will retain for the United States a community of world leading inertial confinement plasma physicists. It is the existence of this community that perhaps more than any other aspect of the NIF project will ensure the long-term technical viability of the United States nuclear arsenal.

The United States is not the only country pursuing laser-driven inertial confinement fusion research with the goal of maintaining a thermonuclear weapons capacity. In France work is under way on a €2.1 billion project known as Laser Mega Joule (LMJ). Like NIF it is intended to be operational by 2010.

France has started with an intermediate project known as the Ligne d'Integration Laser (LIL) under construction by the French Commissariat à l'Énergie Atomique (CEA) at its CESTA site near Bordeaux. The LIL facility briefly held the world record for laser power with a 9.5 kJ laser pulse in April 2003 [50]. By the end of the following month the United States had reclaimed the record with a 10.4 kJ pulse at NIF [51].

The French LIL prototype will be completed with eight converging beams while the full-blown LMJ is planned to have a massive 240. Both the US NIF and the French LMJ teams aim to produce 1.8 MJ pulses using similar optical arrangements and hohlraum geometries [52].

The United Kingdom, however, regarded such large-scale laser systems for nuclear weapons research as unaffordable and decided not to construct such a facility in Britain. In June 1999 in answer to a parliamentary question the British Defence Secretary George Robertson said 'I announced in the Strategic Defence Review that for as long as the United Kingdom has nuclear forces we will ensure that we have a robust capability to underwrite the integrity of our nuclear warheads without recourse to nuclear testing. As part of that I have approved investment in the US National Ignition Facility (NIF). This will guarantee the United Kingdom access to a high-powered laser, which is a key element of our stewardship programme. Participation in the NIF will be a joint venture under the auspices of the 1958 UK/US Mutual Defence Agreement.'

Whether the British nuclear weapons scientists of AWE Aldermaston will also be able to help push forward UK involvement in ICF approaches to electricity production is not clear to the outside observer. What is anticipated is that UK civilian researchers will be able to pursue experiments at the NIF. The risk is perhaps that by contracting with the US rather than constructing a domestic resource the important synergies necessary for ICF electricity development might be missed by the UK.

It has been the existence of defence-oriented policy goals that ensures the vitality of ICF approaches to fusion irrespective of the eventual likelihood of the techniques having anything to offer electricity generation. This book is concerned for the future of nuclear electricity energy production. Looking to the long-term it would reckless to exclude the possibility that one day the power of the hydrogen bomb could be harnessed peacefully, just as the power of the atom bomb was harnessed in the 1950s. If this is to be achieved it is not impossible that this may be done using the physics of thermonuclear weapons (ICF) rather than the utterly civilian physics of magnetic confinement fusion.

III.11.2 The *Z*-pinch

While the National Ignition Facility (NIF) has been beset by technical difficulties and management turmoil, another approach to inertial confinement fusion has been making steady and impressive progress. The principle of the *Z*-pinch approach to fusion is that the radiation pressure required to compress the fusion pellet need not be applied directly. The conventional approach to inertial confinement is to use laser beams to implode the fusion pellet—this is the approach being adopted at NIF. Interestingly the Z-pinch approach uses plasma physics at two stages in the fusion process. The first plasma produces the x-rays fundamental to the process, while the second involves plasma physics more conventionally as the x-rays transform the fuel pellet into the plasma needed for nuclear fusion to occur.

In the Z-pinch experiments the hohlraum is a small metal cylinder built with very great care as to its symmetry and regularity. The hohlraum acts as a uniform and symmetric source of x-rays for the inertial implosion of the fusion pellet. The implosion occurs when the x-rays heat the special coating around the pellet known as the ablator, described previously. The rapid emission of particles from the outer surface of the ablator causes a symmetric and sudden compression of the fusion pellet. It is this extreme compression and the sudden rise in temperature that causes nuclear fusion to occur.

The hohlraum produces the x-rays required as a result of its being heated suddenly to 2–3 million degrees [44]. Interestingly while the whole purpose of the hohlraum is to produce x-rays, in the Z-pinch concept, x-rays heat the hohlraum itself. The Z-pinch makes use of a separate plasma physics process in order to generate the x-rays needed to heat the hohlraum. However, unlike the D–T plasmas considered for the fusion process, the x-ray producing plasma of the Z-pinch is most effective if composed of heavier ions.

The Z of Z-pinch refers to the vertical axis of a three dimensional Cartesian space. Just as the X and Y axes define a horizontal plane, so the Z axis emerges vertically in a third dimension. The Z-pinch concept relies on a cylindrical shell of plasma oriented along the Z axis (see figure III.11.4).

Earlier we referred to the large peak power levels associated with the Sandia National Laboratory Z-machine (see figure III.11.5). This is the largest Z-pinch experiment and it passes a massive 20 MA of electrical current through the cylindrical plasma in a fraction of one millionth of a second.

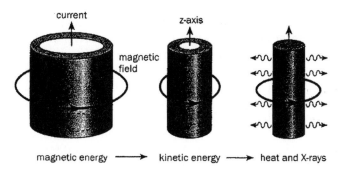

Figure III.11.4. X-ray production in a dynamic Z-pinch (source: Chittenden [44]).

Figure III.11.5. Electrical discharges cover the Sandia National Laboratory Z-machine during a shot (photo courtesy of Sandia National Laboratories).

The result of this massive current is to generate a circular magnetic field in the horizontal X–Y plane wrapping around and through the plasma. This magnetic field causes the plasma to move at right angles to the field, i.e. in a radial direction. The physics is such that the direction of movement of the plasma is towards the Z axis. That is, the cylindrical plasma becomes compressed or 'pinched'. Because of the very rapid pulse of electrical current, the resulting inward collapse of the plasma is also very rapid and the plasma achieves significant kinetic energy. If the ions for the plasma are chosen properly once the ions all collide with each other at high energy near the Z axis then much of the kinetic energy is converted to x-ray radiation. For instance, with a tungsten plasma approximately 20% of the electrical energy discharged through the plasma can be converted into x-rays [44]. When the Z-pinch concept was first developed, it was planned that the cylindrical plasma would be prepared in advance of the electrical discharge. While some fundamental principles were demonstrated, the x-ray yields achieved were far too low for the Z-pinch concept to seem realistic as a route to nuclear fusion. A breakthrough was made, however, with the realization that the cylindrical plasma could be created during the electrical discharge itself. The concept of the wire array Z-pinch involves the prior preparation of a delicate and very precise fine wire cage. It is along this cylindrical cage that the mega-amps of electrical current is discharged, causing the wire cage to be instantaneously vaporized and ionized into a plasma. In 1999 *Science* magazine quoted Robert Heeter of Lawrence Livermore National Laboratory on the power of the Z-machine: 'When you set off the Z the whole building shakes. You can feel it for a few hundred metres in any direction' [53].

The x-rays produced from the Z-pinch then strike the hohlraum wall which is heated to several million degrees and which in turn radiates new

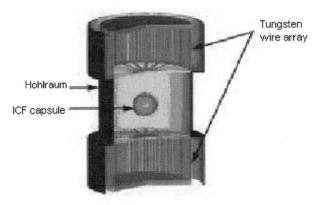

Figure III.11.6. A schematic of a hohlraum/wire array assembly surrounding a fusion fuel pellet and its ablater. This geometry has two Z-pinch wire arrays at the top and bottom of the device with the central cylindrical walls being devoted to the hohlraum. This geometry is proposed for Sandia National Laboratory's proposed machine, the X-1 [44].

x-rays in a symmetric way to strike uniformly the surface of the ablator surrounding the fusion pellet—causing it to implode and the D–T fuel inside to fuse.

Key experiments on wire arrays were performed at Sandia National Laboratory in New Mexico in 1995 and massive x-ray pulses were observed. The Z-pinch at last looked like a realistic prospect for fusion power generation and the researchers started their plans to build the Z-machine.

Figure III.11.6 illustrates a possible structure for the central workings of a Z-pinch machine designed to achieve fusion ignition. Z-pinch wire arrays are located at the top and bottom of the device and the central portion is devoted to the hohlraum and the ablator/fuel fusion pellet. While believed to be workable, such a geometry is intrinsically inefficient, as relatively little of the x-ray power developed at the Z-pinch wire arrays would be successfully transformed into uniform x-rays from the hohlraum wall. This inefficiency means that any device based on this geometry, such as the X-1 machine proposed by researchers at Sandia National Laboratory, would require enormous Z-pinch x-ray production. Even though x-ray production from a Z-pinch is believed to scale as the square of the electrical current applied, it is still the case that the proposed X-1 device would require a massive 60 MA peak current. Interestingly, the X-1 is named after the brightest x-ray source observed, the star Cygnus X-1, and not because it is an example of the particular geometric variant known as an 'X-pinch'.

The X-pinch is a variant of the Z-pinch concept in which the wire array that forms the x-ray producing plasma is not constructed in the form of a cylinder, but rather in the form of a twisted pair of wires forming a cross. Experiments on X-pinches performed on the Mega-Ampere Generator for

Plasma Implosion Experiments (MAGPIE) facility at Imperial College in London have done much to improve understanding of x-ray production from pinched plasmas.

Conventionally the hohlraum of a Z-pinch experiment is a separate structure, as shown in figure III.11.6. One way of overcoming the x-ray conversion inefficiencies is to regard the metal ion plasma of the Z-pinch as the hohlraum itself. This is the dynamic Z-pinch which, while more energetically efficient, it is far more difficult to construct with the symmetry and uniformity required for uniform x-ray irradiation of the plasma fuel pellet.

In recent years Z-pinch physics has advanced a great deal using dedicated budgets far lower than those being applied to either tokamaks such as ITER or to laser systems such as the NIF. Manifestly Z-pinch systems are far from direct utility, particularly in the area of electricity production. The Z-pinch is clearly a nuclear renaissance technology for the long term as much research and development work is still required. In the area of inertial confined approaches to electricity production, however, the Z-pinch concept does have one clear benefit over the laser-driven approach adopted for the NIF. The benefit is that the Z-pinch fuel pellet can be far larger than the fuel pellet that can be accommodated in a laser-driven system [44]. This means that the maximum fusion energy output from a single Z-pinch shot can be far larger than from a single laser-driven event. It is expected that the maximum fusion output a Z-pinch machine might achieve would be 1.2 GJ, which is roughly 80 times that attainable in laser-driven systems [44]. If in the distant future the fusion assemblies of a Z-pinch machine could be prefabricated in mass production then a real industrial-scale power station could be produced on Z-pinch principles. As with laser-driven ICF, a Z-pinch power station would need to be repeatedly fuelled. For a Z-pinch device, however, the thousands of individual fuel elements would be more complex to manufacture than in the laser-driven case considered earlier. Each fusion pulse would require a manufactured assembly comprising D–T fuel, ablator, hohlraum and wire array in a system not unlike bullets being fed into a machine gun. The idea of mass-producing thousands or millions of high reliability fuel assemblies each of which is currently the result of a highly skilled customized craft process is not so fanciful. There are numerous examples from inexpensive digital watches to soda can ring-pulls of inexpensive high precision and high reliability items that would cost enormous amounts to produce as one-off individual items.

III.11.3   Ion beam inertial confinement fusion

Ion beam approaches to inertial fusion seem increasingly unlikely to form part of any nuclear renaissance. Both Europe and the United States had significant heavy ion experiments devoted to nuclear fusion in the 1980s and early 1990s, including those at GSI Darmstadt in Germany and at Sandia National Laboratory in New Mexico in the United States. The aim of these experiments was to use converging beams of ions to compress a fusion pellet. Once again the issues of uniformity of irradiation and uniformity of pellet compression proved problematic. In the United States the PBFAII machine was redesigned and re-commissioned as the Z-machine discussed in the preceding section. US interest in heavy ion ICF continues with the Heavy Ion Virtual National Lab. This collaborative venture was established between the Lawrence Berkeley National Laboratory, the Lawrence Livermore National Laboratory and the Princeton Plasma Physics Laboratory in 1999. Currently emphasis is being placed on a series of small-scale experiments into intense beam physics. The team's ambition is to construct an Integrated Beam Experiment (IBX), a single-beam induction accelerator combining beam injection, electrostatic and magnetic transport, acceleration, steering, and chamber transport [54]. In Germany the 'HIDIF' experiments at GSI were terminated in 1998 [55].

Heavy ion convergent beam physics continues with major facilities such as the Relativistic Heavy Ion Collider (RHIC) at Brookhaven National Laboratory on Long Island, New York. These experimental facilities are, however, not constructed with a view to nuclear fusion research or to electricity production but rather with a view to better understanding fundamental nuclear-particle physics, such as the quark-gluon plasma of the early universe.

Researchers at the University of Wisconsin and Sandia National Laboratories have developed conceptual designs for light-ion-driven fusion power plants based upon experiments using lithium ions. As with heavy-ion-driven ICF there seems little likelihood of electricity production from such approaches on timescales more rapid than those of magnetically confined fusion or even the laser-driven and Z-pinch ICF concepts discussed above [45].

For a long time heavy ion inertial confinement seemed to have more promise than the, now dominant, technique of laser-driven inertial confinement. In the early years it seemed unlikely that lasers would be able to generate sufficient momentum to implode fusion pellets effectively. Heavy ions, on the other hand, would easily achieve the magnitude of compressive forces required. The process by which laser inertial confinement came to dominance is really a consequence of greater than expected improvements in laser physics and optical engineering. It seems likely that if inertial confinement fusion is to develop into an effective fusion power plant it will indeed use the momentum of light either from ultraviolet lasers or from the x-rays of a Z-pinch.

Part III references

[1] European Commission, *Sixth Framework Programme, FP6 Budget* (as of July 2004: http://www.cordis.lu/fp6/budget.htm)

[2] Henderson R 2003 *Developing and Managing a Successful Technology and Product Strategy* (Cambridge, MA: MIT Sloan Learning Resources, MIT)

[3] Taleyarkhan R P, West C D, Cho J S, Lahey Jr R T, Nigmatulin R I, and Block R C 2002 'Evidence for nuclear emissions during acoustic cavitation' *Science* **295** 1868–1873

[4] Eastman T E, Plasmas International, *Perspectives on Plasmas* (as of July 2004: http://www.plasmas.org/industrial)

[5] Braams C M and Stott P E 2002 *Nuclear Fusion: Half a Century of Magnetic Confinement Fusion Research* (Bristol: Institute of Physics Publishing)

[6] Hill J W and Petrucci R H, *General Chemistry—An Integrated Approach*, 3rd edition, Companion Website, Prentice Hall, chapter 19 (as of July 2004: http://cwx.prenhall.com/bookbind/pubbooks/hillchem3/)

[7] Wikimedia Foundation, *Wikipedia, the Free Encyclopedia*, St Petersburg, Florida (as of July 2004: http://en.wikipedia.org/wiki/Nuclear_fusion)

[8] Braams C M and Stott P E, op cit. pp 26–27

[9] Braams C M and Stott P E, ibid. pp 210–214

[10] Glanz J 1996 'Turbulence may sink titanic reactor' *Science* **274**(5293) 1600

[11] Pownall I 1997 'Collaborative development of hot fusion technology policies: strategic issues' *Technology Analysis and Strategic Management* **9**(2) 193–212

[12] Miller A I 2001 'Heavy water: a manufacturers' guide for the hydrogen century' Canadian Nuclear Society Bulletin 22(1)

[13] Ferguson C D 1999 'Tritium: TVA gets the nod' *The Bulletin of the Atomic Scientists* **55**(2) 12–14

[14] Glanz J 1999 'Common ground for fusion' *Science* **285** 820–821

[15] American Institute of Physics 2004 'Physics update' *Physics Today* **57**(1) 9

[16] Arthur D. Little Ltd. for Department of Trade and Industry UK, *Evaluation of the UK Fusion Programme, Part A*, November 2001

[17] Bechmann G, Gloede F and Lessmann E 2001 'International power supply policy and the globalisation of research: the example of fusion research' *Fusion Engineering and Design* **58/59** 1091–1095

[18] Max-Planck-Institut für Plasmaphysik, press release, *USA and China join ITER International Cooperation*, Garching, Germany, 3 February 2003

[19] Dean S O 1995 'Applications of plasma and fusion research' *Journal of Fusion Energy* **14**(2) 251–279

[20] Risken H 1996 'The Fokker–Planck equation methods of solution and applications' *Springer Series in Synergetics*, vol 18

[21] *Hybrids Promise, Problems, and Progress*, Proceedings of US/USSR Workshop on Fusion–Fission Reactors, Moscow, USSR, March 1977

[22] Manheimer W 1999 'Back to the future: the historical, scientific, naval, and environmental case for fission fusion' *Fusion Technology* **36** 1–15

[23] Manheimer W 2000 'Can a return to the fusion hybrid help both the fission and fusion programs?' *APS Forum on Physics and Society, American Physical Society* **29**(3)

[24] Wagner L M and the ARIES Team 2000 'Assessing a new direction for fusion' *Fusion Engineering and Design* **48** 467–472

[25] Barnett C 1996 *The Audit of War* part 3 (London: PAN Books)

[26] Braams C M and Stott P E, op. cit. p 15

[27] Braams C M and Stott P E, ibid. pp 191–192

[28] Hoang G T and Jacquinot J 2004 'Controlled fusion: the next step' *Physics World* **17**(1) 21–25

[29] Najmabadi F, *Fusion Power Plants: Goals and Technological Challenges*, presentation at 31st IECEC Conference, Washington DC, 11–16 August 1996

[30] Federici G, Loarte A, and Strohmayer G 2003 'Assessment of erosion of the ITER divertor targets during type I ELMs' *Plasma Physics and Controlled Fusion* **45**(9) 1523–1547

[31] Koch G 'Bush right to heed call to the stars' *The Stanford Daily*, 12 January 2004

[32] Muroga T, Gasparotto M and Zinkle S J 2002 'Overview of materials research for fusion reactors' *Fusion Engineering and Design* **61/62** 13–25

[33] European Parliament, *Workshop on Operational Requirements of a Commercial Fusion Reactor*, Brussels, 30 September 1997

[34] Patterson W 1999 *Transforming Electricity* (London: Earthscan)

[35] Paméla J and Solano E R 2003 *From JET to ITER: Preparing the Next Step in Fusion Research*, JET EFDA Presentations

[36] Ioffe Institute 2001 *Globus-M Tokamak Experiment*, Russian Academy of Science (as of July 2004: http://www.ioffe.ru/HTPPL/)

[37] UKAEA Culham 2004 *MAST: The Spherical Tokamak at UKAEA Culham* (as of July 2004: http://www.fusion.org.uk/mast/)

[38] Princeton Plasma Physics Laboratory 2003 *National Spherical Torus Experiment* (as of July 2004: http://nstx.pppl.gov/)

[39] Neumeyer C, Heitzenroeder P, Kessel C, Ono M, Peng M, Schmidt J, Woolley R and Zatz I 2003 'Next step spherical torus design studies' *Fusion Engineering and Design* **66–68** 139–145

[40] Ono M *et al* 2003 'Progress towards high performance, steady-state spherical torus' *Plasma Physics and Controlled Fusion* **45** A335–A350

[41] Najmabadi F and the ARIES Team 2003 'Spherical torus concept as power plant—the ARIES-ST study' *Fusion Engineering and Design* **65** 143–164

[42] Braams C M and Stott P E, op. cit. p 18

[43] Braams C M and Stott P E, ibid. p 47

[44] Chittenden J P 2000 'The Z-pinch approach to fusion' *Physics World* May 39–43

[45] Lawrence Livermore National Laboratory, Inertial Fusion Energy: *Opportunity for Fusion Innovation*, January 1997, LLNL ref: UCRL-MI–125743 (as of July 2004: http://www.llnl.gov/nif/library/ife.pdf)

[46] Seife C and Malakoff D 2000 'Will Livermore laser ever burn brightly?' *Science* **289** 1126–1129

[47] Sawicki R, *The NIF Target Chamber: Ready for the Challenge*, Science and Technology Review, Lawrence Livermore National Laboratory, May 2001

[48] Malakoff D 1999 'DOE slams Livermore for hiding NIF problems' *Science* **285** 1647

[49] Heller A, *Target Chamber's Dedication Marks a Giant Milestone*, Science and Technology Review, Lawrence Livermore National Lab, September 1999

[50] Commissariat à l'Énergie Atomique, *L'Actualité du CEA*, 14 May 2003, Paris, France (as of July 2004: http://www.cea.fr/fr/actualites/articles.asp?orig = com&id = 425)

[51] National Ignition Facility Program, *NIF Early Light Effort*, Lawrence Livermore National Laboratory, June 2003 (as of July 2004: http://www.llnl.gov/nif/project/news_nel1.html)

[52] Optics.org, *Laser Fusion Achieves Prototype Milestone*, originally from: Opto & Laser Europe September 2002 (as of July 2004: http://optics.org/articles/ole/7/8/3)

[53] Watson A 1999 'Z mimics x-rays from neutron star' *Science* **286**(5447) 2059
[54] US Heavy-Ion Fusion Program, 2004, Lawrence Berkeley National Laboratory website (as of July 2004: http//hif.lbl.gov)
[55] Hofmann I 2001 'Heavy ion inertial fusion in Europe' *Nuclear Instruments and Methods in Physics Research A* **464** 24–32

Afterword

This book provides an overview of issues facing the future of nuclear power. Within it the central issue is: will there be an extensive programme of new nuclear power plant construction in Europe and North America? Frankly, this author does not know the answer to that question, but on balance I would speculate that such a renaissance in nuclear power does indeed seem likely.

Our approach has been based on the premise that there are three fundamental policy issues that will shape any nuclear renaissance. They are the economics of nuclear power, the environmental factors surrounding nuclear energy and the security of electricity supplies. This author believes that these three factors are of roughly equal importance. They fluctuate in emphasis with time and, at any one time, different countries may have different dominant concerns. Fundamentally, however, all three factors are of central importance to the future of nuclear power.

We have examined the economic costs of generating electricity from nuclear power. It seems probable that the lifetime costs of modern nuclear power plants are indeed competitive. This assessment is further favoured by the significant increases in oil prices during the first half of 2004. The most challenging economic issue is that nuclear power is extremely capital intensive. This has made it unattractive to investors in liberalized electricity markets, but as these markets mature it seems likely that project finance for nuclear power plant development will become easier.

Many would argue that at present the overwhelming policy issue shaping nuclear power is its demonstrated ability to produce large amounts of base load electricity without the production of harmful greenhouse gases. Nuclear power has much to contribute to the minimization of global warming. Many proponents of nuclear power believe that the intermittency of wind power and most other renewable energy sources means that they will never be sufficiently reliable to serve the electricity needs of a modern industrialized economy. This author does not share such pessimism. I believe that, for instance, wind power can and should make a significant contribution to UK electricity generation. However, no single source of electricity (wind, nuclear, gas, coal etc.) should ever provide more than half of the UK's electricity supply, even on a warm and breezy summer afternoon. Maintaining the contribution of nuclear power, increasing the contribution of renewables, and working hard to improve the environmental performance of fossil fuels should all form part of a balanced attempt to reduce carbon dioxide emissions produced by electricity generation.

A balanced portfolio approach to electricity generation also has direct benefits in the important area of security of supply. Nuclear power has much to contribute to security of electricity supplies. Given the inevitable depletion of fossil fuel resources and the increasing cost of securing new sources as reserves diminish, security of supply can only rise in importance for electricity generation. It may even be the case that this process of the fossil fuel endgame has already begun.

The real economics of electricity go far beyond the cost of building power stations and sourcing the fuel. The electricity system is a critical infrastructure in every industrialized country. An unreliable system, with extreme price volatility and systemic vulnerabilities, can be damaging at a macroeconomic level.

When we consider the three main policy issues underlying this book it seems probable that nuclear power does indeed have a future. This author believes that nuclear power deserves to be considered on its merits. It has no manifest right to contribute to power generation, but nor is it intrinsically evil.

I regard commercial nuclear power as a beneficial but not an essential technology. With some lingering doubts, I do regard fission-based nuclear weapons as having been a net good, because I believe they have reduced conflict. I note that for the most part they have so far been available to responsible governments only. In this book our consideration of the three major policy factors shaping the future of civil nuclear power has meant that we have largely avoided a key topic that represents the most negative aspect of a nuclear future—the risk of an eventual proliferation of nuclear weapons to rogue states or even terrorist groups.

This author does not subscribe to the view that technologies have politics [1]. For instance, I do not believe that the separation of plutonium must eventually lead to a surveillance society with reduced civil liberties, but I do have concerns about such things. I do believe, however, that new technologies can raise new policy issues and new risks to society. Technologies can prompt us to make choices concerning the future direction of our society.

Mankind's use of nuclear energy is only approximately 60 years old. In these years we have risked nuclear war and nuclear terrorism. As I write these words, mercifully, nuclear power has not been used to kill since 1945. In his book *Our Final Century* Sir Martin Rees, the Astronomer Royal, has recently reminded us how fragile our civilization really is [2]. I share his concerns and I believe that the long-term risks of malevolent use of nuclear energy are sufficiently great to cause me to pause, before recommending any nuclear renaissance.

In saying that I have worries concerning the deliberate misuse of nuclear power over the long-term I do not mean to imply that I am worried about the weeks or years immediately ahead. I mean to say that I am worried about the next hundred years and the next several hundred years. How will society manage the risks of the ever-spreading availability of fissile materials?

Given these remote but genuine risks, do I wish that civil nuclear power had never been developed? To that my answer is clear—I am glad that nuclear power was developed and I believe it to have made major positive contributions to society. Furthermore, despite the military uses made of nuclear fission, I am also glad that the physical phenomena involved were discovered in the late 1930s.

My concerns for the long-term future, however, lead me to an unusual (and wholly impractical) conclusion. I actually wish that the laws of physics were different. I wish that nuclear fission did not exist. To be clear, I am not saying that I wish nuclear fission had never been discovered, I am saying that I wish that the physical phenomenon did not exist.

The secrets of nuclear fission are sometimes described as a Pandora's Box that should never have been opened. The term 'Pandora's Box' refers to an Ancient Greek myth in which Pandora was the first woman. She was created by Hephaestus and Zeus as revenge upon Prometheus who had stolen fire from heaven. Pandora married Prometheus' brother Epimetheus and her dowry to him was a mysterious and attractive box. Once Prometheus opened the box, it spewed forth evil into the world. The evil could never be gathered up and returned to the box. Once the evil had been released, humanity had to live with it forever. Key to the Pandora's Box analogy is a presumption that nuclear fission is evil. There is also a second implication: that humanity can never un-learn the secrets of nuclear fission. It is primarily the latter aspect of the metaphor that secures this author's attention.

Perhaps the early research into nuclear fission was indeed a period in which mankind toyed with a Pandora's Box. Perhaps the pressures of World War II, and the establishment of the Manhattan Project dedicated to creating a fission weapon, represented the actual opening of the box. If one holds such view, where then does civil nuclear power fit in?

Many with a negative opinion of nuclear power dwell upon its links with nuclear weapons and the military industrial complex and see it all as part of the same spreading evil. On the side of such thinking is the fact that it does indeed seem likely that without the initial push for nuclear weapons in the 1940s and 1950s, it would not have been possible for civil nuclear power to develop in the way that it did in the late 20th century.

I do not subscribe to the view that nuclear fission is a true Pandora's Box, because the implied evil nature of nuclear fission in the metaphor is too absolute for my taste. What emerged from the discovery of nuclear fission has so far proved to be mostly positive and benign, but, as noted above, I do have concerns for the long-term future as nuclear knowledge can only spread.

At the risk of taking too far a metaphor with which I disagree, I should say that if the 'box' existed then mankind simply had to open it. In history it was the allies led by the United States that opened the box of applied nuclear fission. If the box exists, then it is inevitable that eventually someone would have taken the fateful step. Better that a democratic super-power did it than some ill-constituted group of criminals. In writing this I am conscious that in an alternative 20th century Nazi Germany might have opened the box first. There are risks and luck in all of this.

If the box exists then we must open it, but would it not be better for the long-term if the box did not exist at all? In such a world we would be truly free forever from the threat of nuclear attack. A world without nuclear fission would have had a rather different history in the past 70 years. It is interesting to consider briefly some of the possible attributes of such a world.

In the military field: without nuclear fission World War II in the Far East would probably have lingered on with many more deaths. The Soviet Union might have captured more of Japan than simply the small Kuril Islands. A communist Japan is unlikely to have pioneered developments in computing and electronics in the late 20th century. The massive loss of life at Hiroshima and Nagasaki in Japan in the nuclear bombings of 1945 needs to be set beside the similarly horrific death tolls from the conventional bombing of cities such as Tokyo, Hamburg and Dresden. The Cold War would surely have occurred without nuclear weapons, and it would most probably still have relied on ballistic missile technology. Whether chemical and biological weapons would have been able to achieve their current pariah status in the absence of nuclear weapons is an interesting issue given their compatibility with missile technology. It might be argued that poison gas had achieved such a status after World War I, but I leave such matters to the historians. It is widely believed that nuclear weapons kept the peace during the second half of the 20th century, albeit at the price of significant risks to global safety. It is interesting ask to what extent deterrence between the great powers in the Cold War actually required the existence of nuclear weapons, given the harm the two super-powers could have done to each other by other means.

In addition to the contribution of nuclear weapons to Cold War deterrence, one must also consider the contribution of nuclear fission to the development of the ballistic missile submarine. Submarines powered by nuclear reactors made possible the assuredness of mutually assured destruction that underpinned deterrence during the Cold War. Nuclear power made the strategic nuclear weapons submarines almost undetectable and so made it evident that any country possessing such a technology could retaliate with it in response to a 'bolt from the blue' attack on the motherland.

It is interesting to ask how far nuclear fusion research would have progressed in the absence of nuclear fission. Given the impossibility of creating a thermonuclear weapon in the 1950s, it seems likely that even the entirely civilian magnetic confinement approach to fusion would not have emerged at anything like the pace that has actually been seen. If nuclear fusion power is to drive humanity's development in the late 21st century and beyond, then perhaps this is an example of a benefit made possible by the existence of nuclear fission.

Commercial nuclear power for electricity generation was operating on a large scale for approximately 20 years before the first hints of the problems of global climate change became available. It is unlikely that the science of global warming would have been understood much earlier even in a scenario where the carbon dioxide emissions into the atmosphere had occurred more rapidly. However, a world without nuclear fission would surely have faced worse threats to the climate than we actually face today.

It is likely that electricity would have been cheaper and would have been privatized sooner in the major industrialized economies if nuclear fission had not existed. This would have improved national economic competitiveness and probably led to higher levels of wealth for ordinary people in those countries.

In imagining a world without fission one is led to ask if is it possible to have the range of radioactive processes required by medicine and industry without the existence of nuclear fission. One is even led to ask if nuclear fission is a requirement for life on earth. The answer to that at least is probably 'no'. Nuclear fission is of course a natural phenomenon. The clearest illustration of this is the existence of natural nuclear fission reactors in Oklo, Gabon, West Africa [3]. However, it is virtually certain that these natural reactors have had absolutely no effect on human development. One hypothesized nuclear reactor might have a greater impact on life on earth. J M Herndon and D F Hollenbach suggest that there are, or have been, natural fast reactors deep in the earth [4]. They point to data concerning geomagnetic field reversals and the ratio of the two helium isotopes He-3 and He-4 as evidence in favour of their controversial hypothesis. If their hypothesis is correct and such fission reactions contribute significantly to the temperature of the deep earth, then there is a remote possibility that the existence of the phenomenon of nuclear fission could have affected humanity prior to the discovery of the phenomenon in the late 1930s. The wish that the phenomenon of nuclear fission does not exist can be sustained only if there are indeed no natural benefits of nuclear fission.

One area in which nuclear fission has profoundly affected people is that of public anxiety. This ranges from disquiet for those uncomfortable with nearby nuclear facilities, through to serious psychological problems caused by proximity to nuclear accidents. This anxiety is one of the central realities of nuclear power and is something that the industry should have managed much more carefully.

The technology of the thermonuclear fusion weapon is one that might have been arrested by early unilateral decisions of the leading powers. In my view this technology is of no significant military utility, especially as missile targeting has become more precise. It seems likely that all the great powers would have realized its limitations sooner if the United States or the Soviet Union had taken a stand against the technology at an early stage.

For those concerned with weapons of mass destruction and nuclear proliferation the increasing availability of dirty bombs and fission weapons is, of course, a prime concern. Even the worst-case scenario (a terrorist group detonates a fission weapon in a major city) we can be sure that civilization broadly as we know it could survive. The real fear must be that one day, perhaps many decades from now, a terrorist group or a failed state will develop a two-stage thermonuclear weapon. That would be the new doomsday scenario, for the world could not sustain too many attacks on major cities with such weapons. Whatever the endgame, we could be sure life would be changed irrevocably. It is true that one hears little mention of the long-term risk of renegade groups obtaining thermonuclear weapons. However, in 2003 *The Bulletin of the Atomic Scientists* published a telling statistic. For the big five nuclear powers, the average delay from first fission bomb to first multistage thermonuclear bomb was only 71 months [5]. If a nation state can make the transition in less than six years, how many decades will it be before such technologies are available to the truly irresponsible?

If the physical phenomenon of nuclear fission did not exist, it also seems most likely that the world would be spared for eternity from the greater risk arising from two-stage thermonuclear weapons.

In this Afterword I have largely concerned myself with the hypothetical. Would the world be a better place if nuclear fission did not exist? While this author is inclined to agree with such a view, the reality of course is that the physics of nuclear fission does exist and has always existed. Given this reality, this author supports the development of commercial nuclear power in the years after World War II. The continued threat of nuclear proliferation forces the great powers to remain engaged with the issues of nuclear fission. Perhaps one way to do this would be to maintain, in perpetuity, a small cohort of technical experts able to advise on and respond to any threats. However, in the absence of a continuing commercial nuclear energy programme these experts would eventually become an elite detached from everyday life. They would soon merit the memorable label first coined by Thomas Sebeok—*the atomic priesthood*. It is my impression that it is preferable that ordinary people remain in touch with these issues and remain sensitive to both the benefits and disadvantages of nuclear technology.

Nuclear engineers and policy makers should be ordinary citizens following ordinary career paths. Ensuring that nuclear knowledge is part of the fabric of society, rather than a marginal dark art, is best assured by the continuation of the commercial nuclear energy business. Of course in this scenario more people will know the secrets of harnessing nuclear fission but, if handled carefully, this need not speed the flow of dangerous knowledge to those who seek to harm others. A wider societal engagement with these issues could bring with it a concomitant awareness and vigilance that would provide long-term protection against the greater risks of complacency and ignorance.

Any attempt to move into the future without a nuclear renaissance seems dangerously utopian, given the unfortunate realities of the world in which we live. At present the two greatest threats to the developed world seem to be climate change and the weaknesses of homeland security. In this Afterword I have raised issues concerning the long-term security of our world, and elsewhere in the book I have considered the environmental credentials of nuclear power in depth.

On balance it would seem prudent for the developed world to maintain a civil nuclear power industry on at least its current scale. This will help stabilize the global climate and provide us with the widest base of knowledge and capacity with which to challenge the long-term risks of nuclear related terrorism.

References

[1] Winner L 1986 *The Whale and the Reactor* (University of Chicago Press) pp 19–39
[2] Rees M 2003 *Our Final Century* (London: Heinemann)
[3] *Natural Fossil Fission Reactors*, website of Curtin University (as of July 2004: http://www.curtin.edu.au/curtin/centre/waisrc/OKLO/)
[4] Hollenbach D F and Herndon J M 2001 *Proceedings of the National Academy of Sciences* **98**(20) 11085–11090
[5] Norris R S and Kristensen H M 2003 *Bulletin of the Atomic Scientists* Sept/Oct, p 71

Index